GEORGE R.

GEORGE, by the Grace of God, King of *Great Britain, France* and *Ireland*, Defender of the Faith, &c. To all to whom these Presents shall come, Greeting. Whereas Our Trusty and Well-beloved *Samuel Buckley*, Citizen and Stationer of Our City of *London*, has humbly represented unto Us, that he is now printing a Book written by Our Trusty and Well-beloved *Humphrey Bland*, Esq; Lieutenant-Colonel of Our own Regiment of Horse, Intitled, *A Treatise of military Discipline; in which is laid down and explained the Duty of the Officer and Soldier, through the several Branches of the Service*: And whereas the said *Samuel Buckley* has informed Us, that he has been at a great Expence in carrying on the said Work, and that the sole Right and Title of the Copy of the said Book is vested in the said *Samuel Buckley*; he has therefore humbly besought Us to grant him Our Royal Privilege and Licence for the sole Printing and Publishing thereof, for the Term of Fourteen Years: We, being willing to give Encouragement to so useful a Work, are pleased to condescend to his Request, and do therefore hereby, so far as may be agreeable to the Statute in that Behalf made and provided, grant unto the said *Samuel Buckley* Our Royal Licence for the sole Printing and Publishing the said Book, for and during the Term of Fourteen Years, to be computed from the Day of the Date hereof, strictly prohibiting all Our Subjects within Our Kingdoms and Dominions to reprint or abridge the same, either in the like or in any other Volume or Volumes whatsoever; or to import, buy, vend, utter or distribute any Copies of the same, or any Part thereof, reprinted beyond the Seas, within the said Term of Fourteen Years, without the Consent and Approbation of the said *Samuel Buckley*, his Heirs, Executors, and Assigns, by Writing under his or their Hands and Seals first had and obtained, as they and every of them offending herein will answer the contrary at their Perils. Whereof the Master, Wardens, and Company of Stationers of Our City of *London*, Commissioners and other Officers of our Customs, and all other Our Officers and Ministers whom it may concern, are to take Notice, that due Obedience be given to Our Pleasure herein signified. Given at Our Court at St. *James*'s the twenty-fourth Day of *April* 1727, in the Thirteenth Year of Our Reign.

By His Majesty's Command.

HOLLES NEWCASTLE.

A

TREATISE
OF
Military Discipline:

In which is Laid down and Explained

The Duty of the Officer and Soldier,

Through the feveaal Branches of the Service.

By *HUMPHREY BLAND*, Efq;
Lieutenant-General of His MAJESTY's Forces.

The NINTH EDITION, Revifed, Corrected, and Altered to the prefent Practice of the Army.

In omni Prœlio non tam Multitudo & Virtus indocta, quam Ars & Exercitium folent præstare Victoriam. Vegetius, Lib. 1.

The Naval & Military Press Ltd

Published by
The Naval & Military Press Ltd
Unit 10 Ridgewood Industrial Park,
Uckfield, East Sussex,
TN22 5QE England
Tel: +44 (0) 1825 749494
Fax: +44 (0) 1825 765701

www.naval-military-press.com
www.military-genealogy.com
www.militarymaproom.com

In reprinting in facsimile from the original, any imperfections are inevitably reproduced and the quality may fall short of modern type and cartographic standards.

TO THE
KING.

SIR,

Most Humbly beg Your MAJESTY's Pardon for laying this Treatise at Your Feet; which I Presume to do upon no other Foundation, than that of my Zeal for Your Service: And I hope, from Your MAJESTY's known

DEDICATION.

known Goodneſs, that You will be pleaſed to excuſe its Errors, ſince the only Motive I had for Writing it, was the Deſire I have every Way to approve myſelf,

May it pleaſe Your moſt Sacred MAJESTY,

Your Majeſty's Moſt Humble,

Moſt Obedient, and

Moſt Devoted,

Subject and Servant,

Humphrey Bland.

THE

PREFACE.

FROM the great Reputation of the British *Arms*, Men would be apt to imagine, that several Treatises, of the Art of War, were to be met with in our Language; but when they come to enquire, they will be strangely surprised to find nothing of this Kind of our native Growth, that I know of, except what has been left by the Earl of Orrery, who wrote fifty Years ago: And though his Writings might have been very instructive at that time, yet so many Improvements have been since made (several Instances of which I could easily produce, were there an Occasion) that they can be but of very little use, at present, to young Officers, for whose Sake, chiefly, this Book is published.

As the Nation has abounded with Officers, whose Services leave no Doubt of their Abilities to perform

PREFACE.

form a Work of this Nature, it must be supposed, that their Indolence, or their imagining no Want of it, was the Reason that nothing of this Kind has hitherto appeared: But considering how few old Officers remain, and that they are diminishing every Day, I hope I shall not be censured for having ventured to commit to Writing the little Knowledge I have acquired in Military Matters, for the Instruction of those who are yet to learn; who, in a little Time longer, if they have no Opportunity of going Abroad, and wanting the Examples of old Officers to guide them, may not think it useless to have Recourse to this Account of their Duty, however imperfect it may be.

I am well aware how little Thanks some People may think I deserve for this Attempt; and am very sensible that Standing Armies, and consequently the modelling of them, are little relished in Time of Peace in this Kingdom: And, I must own, as much a Soldier as I am, not to be desired, if we can preserve our King, our Trade, (and, I had almost said, our Civil Government) without them: But such is the State of the World at present, that, I doubt, it will be in vain to flatter ourselves with the Hopes of opposing our Militia (whatever Hearts they may have in their Bellies) to Troops who have the Advantage of Discipline on their Side: Neither are our Treaties, nor Alliances (as appears by Proofs too recent) sufficient without the Ratio Ultima: *And, I believe, it is a true Observation, that, First or Last, Force has been the conclusive Argument of most Treaties,*

and

PREFACE.

and those have been found the best, which have been supported by the best Troops.

I have, throughout my Book, taken every Occasion to inculcate the Necessity of legal Military Subordination. It has been the Practice of all Nations, ancient and modern, even where the People have been blessed with the highest Liberty, never to admit of a military Independance upon their military Superiors: I look upon it as the Band which ties the Whole together, and, without it, all our Rules and Forms to be of no Use. Perhaps it is the great Distinction between regular Troops and Militia, and the Cause why the Former have always had the Advantage over the Latter.

I know it will be objected, that the better our Troops are, the more dangerous they may be; and I confess the Objection to be of so great Weight, that it fits not me to enter into the Question: All that I am pleading for, is, that as long as it shall be thought necessary to keep up a Body of Regular Troops, those Tooops may be put upon such a Foot, as may answer the End, and the sole End, which is expected from them by the Publick. Give me Leave to add, to the Honor of the Officers, that more Instances than One might be given, when, instead of being those servile Tools (which is so much apprehended) of bringing their Country into Slavery, they have behaved themselves with a Zeal for its Liberty, equal, I think, with the best Englishman; if exposing themselves and Families to the Hazard of Starving, may be allowed to be an Instance of it.

THE EDITOR's Preface.

GENERAL BLAND's Treatise of Military Discipline, *originally published in the Year* 1727, *is a Work so universally known and esteemed, as to stand in need of no Encomiums here: But as the Discipline of the English Troops has undergone great Changes since that Time, it was thought, that a Revisal of it, in order to expunge such Parts as were become obsolete, and, by substituting the new, to adapt it to the present Practice of the Army, would be an Undertaking both acceptable and useful to the Public. This consideration induced the* EDITOR *to assume the Task; and, as his principal Motive was to serve the Officers in general, but especially such as have not been long in the Service, he flatters himself, they will indulge him with their Approbation.*

Savoy, March 26, 1762.

W. FAUCITT.

CONTENTS.

CHAP I.

Directions for the forming of Battalions, posting of Officers, &c.

ARTICLE I.

Directions to the Officers and Soldiers about parading of Companies Page 1
About the sizing of men, and method of doing it 2
In what manner the Companies are to march to the General Parade ibid.
How the Companies are to draw up in Battalion, with a plan of the same 3
The old way of completing of files rejected, and a new method proposed. ibid.
Directions for the drawing up of the Company of Grenadiers 5
ARTICLE II. *How the Officers take their posts in Battalion* ibid.
How the Serjeants and Drummers are to be posted ibid.
How the Colonel and Lieutenant-Colonel are to post themselves 7
No Officer to be absent without Leave when the Battalion draws out ibid.
ARTICLE III. *Method of sending for the Colours* ibid.
Compliment paid to the Colours 9
Method of lodging the Colours 11
ARTICLE IV. *Distances of ranks and files* ibid.
Rules for the opening of files 12
ARTICLE V. *Method of forming the Battalion six deep* 14
The standing salute preceding the Exercise 15
ARTICLE VI. *The Major and Adjutant to exercise the Regiment on horseback* ibid.

Other

CONTENTS.

Other Officers, when ordered, to exercise it on foot Page 15
Directions to those who exercise the Regiment 16
Instructions for delivering of the words of command ibid.
Position of a Soldier under arms 17

CHAP. II.

The Exercise of a Battalion of foot, with an explanation 18

CHAP. III.

Manual Exercise of the Grenadiers, with an explanation 46

CHAP. IV.

General Rules for wheeling 56

CHAP V.

Of passing in Review.

ARTICLE I. *The Companies to be inspected before the Review* 62
ARTICLE II. *When the Regiment is formed, what is necessary to be done before the General comes to review it* ibid.
Compliment to be paid the General who reviews 63
Directions for the facing of the Regiment 64
Officers to remain in the front, when the General passes along the rear 65
ARTICLE III. *After the Regiment has been reviewed standing, what further ceremonies are commonly performed* ibid.
ARTICLE IV. *How the Colonel and Lieutenant-Colonel are to post themselves during the Exercise* 66

CONTENTS.

The Major to exercise the Regiment on horseback though he should command it   Page 66

When a Captain commands the Regiment, how he is to proceed   67

How the Hautboys, Drummers and Serjeants on the flanks are to post themselves during the Exercise   ibid.

Officer who exercises not to find fault before the General   ibid.

ARTICLE V. *After the Exercise, how they are to march by the General*   ibid.

What grand and sub-divisions are   ibid.

How they are to march by grand and sub-divisions   68

Manner of posting the Officers to them   ibid.

Directions about saluting   69

How they are to pass by in Companies   71

When they have passed by the General, to draw up on their former ground.   72

ARTICLE VI. *When the Regiment marches off from the left, how the Colonel and Lieutenant Colonel are to post themselves.*   ibid.

CHAP. VI.

Consisting of directions for the different firings of the foot.

ARTICLE I. *How the Battalion is to be formed to go through the firing*   74

Telling off the platoons, and what number of files they consist of   ibid.

Officers to command the platoons, and manner of posting them   75

The rest of the Officers fall in the rear, and reason why   76

Platoons divided into several firings, and number each firing commonly consists of   ibid.

Platoons of each placed in different parts of the Battalion, and reasons for it.   77

ARTICLE II. *Manner of dividing the platoons in different parts of the Battalion*   78

Plan

CONTENTS.

Plan of a Battalion told off in eighteen platoons Page 79

Words of command of the platoon Exercise, with an explanation 82

The Major to be exact in the telling off the platoons, and the Officers to be thoroughly informed of their firing and number 83

Method to try if they know it, to prevent mistakes in the execution ibid.

The whole to be informed in what manner they are to fire 84

Whether standing, advancing, or retiring, or all three; and if the platoons of each firing are to fire in their order (and what is meant by their order) or together ibid.

If to proceed by beat of drum, what beatings the Major will use ibid.

Common beatings used, and what is generally performed at each ibid.

ARTICLE III. *Directions to fire standing, which is performed in their order* ibid.

Directions to the Officers who command the platoons in giving the words of command 85

How to go through the firings advancing 86

Directions to fire retiring 87

How to fire by sub-divisions standing 88

How to fire by sub-divisions advancing ibid.

How to fire by sub-divisions retreating 89

How to fire by grand-divisions, standing, advancing, and retreating ibid.

How to fire by the firings, standing, ibid.

How to fire by the firings, advancing 90

How to fire by the firings, retreating ibid.

ARTICLE IV. *Directions for the firing by ranks, with remarks on it* 91

ARTICLE V. *Parapet firing, on what occasion used, with the methods of performing it* 92

ARTICLE VI. *Street-firing, when used, and how performed at Exercise* 97

ARTICLE VII. *Running-fire, or* Feu de joy, *how performed in Camp and Garrison* 98

CHAP.

CONTENTS.

CHAP. VII.

Containing directions how a Battalion of foot is to defend itself when attacked by horse.

ARTICLE I. *Directions for their telling off, that they may act either in Battalion, or in the Square* Page 101
The first plan the most perfect, as being proper for both ibid.
Plan of a Battalion told off in twelve platoons 102
Apprehensions from horse, proceed from the foot not knowing their own strength ibid.
A situation supposed, where they can be only attacked in front 103
Directions for the management of their fire in that situation ibid.
In this case, the front rank to fire with the rest, and reason for it 104
How to judge of real or feint attacks by the disposition of the Enemy 105
Proposal to advance small platoons to disappoint their feint attacks ibid.
Directions to those platoons in advancing and giving their fire 106
Platoons of the first firing to make ready when those advanced do, and when to give their fire ibid.
A situation supposed where one flank is exposed 107
Manner of securing that flank ibid.
ARTICLE II. *Directions for the forming of the Square from three deep* 108
Two plans shewing how it is done
Directions for the reducing of the Square into Battalion 113
ARTICLE III. *Directions for the forming of the Square by four grand divisions, with a plan of the same* 115
How it is reduced into Battalion 118
Remarks on the different ways of forming the Square 120

ARTICLE

CONTENTS.

ARTICLE IV. *Directions for firing and marching in the Square* Page 121
The platoons of each face divided into different firings ibid.
Figures on the inside shew what firings they belong to 122
How they are to fire in their order ibid.
How they are to face in marching towards the different fronts 125
How they are to fire marching towards each front 126

CHAP. VIII.

General rules for the marching of a Battalion, or a Detachment of men, where there is a possibility of meeting with the enemy.

ARTICLE I. *An Officer's character hardly retrievable if surprised without being prepared* 132
Reflections to induce young Officers to the study of the service 133
No military man above the knowledge of the minutest part of the service ibid.
Confirmed by the practice of the Germans, and the late Glorious King William ibid.
ARTICLE II. *Method proposed to march by platoons instead of divisions. Reasons for it* 135
Van and rear-guards, what numbers they generally consist of 136
General instructions to the van and rear-guards, and advantage of a rear-guard ibid.
ARTICLE III. *Directions for the marching of the Regiment, and how the Officers are to march their platoons* 139
Major and Adjutant, their duty in the above case 140
Directions for the passage of a defile ibid.
The Soldiers not to fasten the tent-poles to their firelocks 141
Consequence that happened by this, and other neglects ibid.

Reflections

CONTENTS.

Reflections on the said miscarriage, by way of precaution to others Page 142

ARTICLE IV. *In inclosed countries, to have small parties on the flanks: design and use of them* 143

Precautions in reconnoitring. If cut off, how to make it known. 144

How to be relieved, with directions to those who are sent 145

Young Officers apt to exceed their orders from a mistaken zeal ibid.

Reflections on the consequence of that, and the contrary behaviour ibid.

When the van-guard discovers any troops, necessary precautions to be taken 146

Instructions to the Officers who shall be commanded from the body to attack a partizan-party, for fear of being drawn into an ambuscade ibid.

Necessary precautions of the Commanding Officer, if his march is obstructed ibid.

His orders must determine him in these cases 147

Article V. *Necessary precautions, and disposition of the Regiment, in marching through an open or champain country* ibid.

When proper to march in four grand divisions, and method of doing it upon the march 149

Directions to form the Square from grand divisions on the march ibid.

Necessary precautions on the appearance of the enemy's horse, and not to form the Square, but march in grand divisions, till there is a necessity for it 151

When in danger, not to be sollicitous about their baggage ibid.

Van and rear-guards drawn in, and to be divided into the platoons ibid.

If the enemy retires, or keeps at a distance, to pursue their march ibid.

Directions for reducing the Square into Grand divisions 152

CONTENTS.

CHAP. IX.

General rules for Battalions of foot, when they engage in the line.

ARTICLE I. *Necessary inspection into the arms and ammunition the day before the action* Page 155
Regiments to be as strong as possible when they go on service ibid.
To be told off into platoons, and the Officers divided to them ibid.
Inspection of those Officers, and reason why it is necessary ibid.
A profound silence in marching up to, and during the action with the enemy 156
The Commanding Officer to give the words of command for all movements ibid.
Drummers to regulate their beatings according to the said words of command ibid.
ARTICLE II. *The line to move slow in marching up to the enemy* ibid.
The Battalions to march up close before they fire, and then fall upon the enemy with their bayonets, with the advantage which may arise from it 157
But if deferred how dangerous ibid.
ARTICLE III. *When the enemy gives way, the soldiers not to break in the pursuit of them* 158
No Battalion to advance before the line in pursuit of the enemy ibid.
Grenadiers to advance and fire on them to keep up their terror ibid.
Directions to the said Grenadiers in advancing ibid.
Danger in Battalions advancing before the line 159
ARTICLE IV. *An article of war quoted against the Soldiers leaving their ranks to plunder and pillage, before the enemy are entirely beat* 160
Consequences on its not being punctually obeyed ibid.

Two

CONTENTS.

Two instances related, wherein it appeared, by way of precaution to others — Page 161

ARTICLE V. *Field-Officers to observe what Regiments they are to sustain, or who sustains them* — 163

They are to remark the Cloathing and Colours of their own troops, that they may distinguish them from the enemy — ibid.

ARTICLE VI. *When the enemy have men posted in the front of the line, to annoy yours in marching up, the Grenadiers should be ordered to dislodge them* — ibid.

ARTICLE VII. *When a Battalion is ordered to retire, what precautions are necessary for the performing of it, with the Reasons why they may be ordered* — 164

When the greatest part of the line retires, the rest to do the same — ibid.

The Battalions in the line to act in concert with one another — ibid.

ARTICLE VIII. *When the first line gives way, how they are to retire* — 165

How the Officers are to act on this occasion — ibid.

Danger which the second line runs of being broke by the first — 166

Officers of the first line to use their utmost diligence to prevent it — ibid.

ARTICLE IX. *An expedient offered, in order to prevent it* — ibid.

ARTICLE X. *Directions to the second line, on the first giving way* — 168

If followed, how easy to repulse the enemy — ibid.

Danger which the men of the second line apprehend, on their seeing those of the first give way, more imaginary than real — ibid.

Reasons shewing them to be so, in order to remove them — 169

The second line's moving on briskly, will, in some measure, remove them — ibid.

ARTICLE XI. *A remark, that the private men form their notion of the danger from the appearance of their Officers* — 170

CONTENTS.

To prevent bad impressions, the Officers should appear serene and chearful — Page 170

Which will have a good effect, if they have an opinion of their military capacity — ibid.

Young or unexperienced Officers cannot influence them in the same manner — ibid.

ARTICLE XII. *When a Battalion in the line is pressed for room, what they are to do* — 171

ARTICLE XIII. *If attacked in the line, or in Brigade, by horse, how they are to act* — ibid.

ARTICLE XIV. *The alternate firing, practised by the Dutch, described* — 172

Advantage and disadvantage which may attend it — 173

The Dutch, by the excellency of their discipline, have surmounted the danger — ibid.

Common notion, that the Sang Froid of the Dutch is owing to phlegm — 174

Advanced only to excuse our own neglect of discipline — ibid.

The French look upon us in the same light, though they have had reason to be convinced of the contrary — ibid.

Foreigners have a notion that we cannot be brought into discipline — ibid.

Our neglect of it has produced that notion — 175

CHAP. X.

Duty of the Infantry in garrison.

ARTICLE I. *Upon marching to garrison the Quarter-Master to be sent before* — 176

Directions to him on his taking of quarters. — 177

ARTICLE II. *When the caserns or barracks cannot contain the troops, manner of quartering of them on the inhabitants, which is called cantoning* — ibid.

ARTICLE III. *Directions to the Quarter Master about meeting the Regiment, and what is proper to be done before they enter the town* — ibid.

CONTENTS.

ARTICLE IV. *Ceremony at the barrier before they enter, and how conducted to the parade* Page 179
When drawn up on the parade, the Colonel to wait on the Governor 180
Town-Major to read the general orders of the garrison to the Regiment ibid.
Then to conduct them to their regimental parade, alarm-posts and quarters, where they are dismissed 181

CHAP. XI.

Usual guards in a garrison, and manner of forming the parade.

ARTICLE I. *How the guards are composed* 182
Number who mount, according to the troops in garrison ibid.
Method commonly observed in it ibid.
Number each guard commonly consists of ibid.
The citadel, generally a garrison of its own 183
The guards mount in the morning ibid.
ARTICLE II. *Directions about viewing the men who mount the guard* ibid.
ARTICLE III. *Directions for the beating of the assembly or troop* 184
How the detachments are to draw up on the parade 185
ARTICLE IV. *Manner of forming the guards, and posting the Serjeants to them* ibid.
Officers to draw for their guards, and their names to be entered ibid.
Manner of drawing up the reserve and horse guards 186
When the guards are to exercise together, how performed 187
How the guards are to march off 188
Horse guards to march last. Reason supposed for that custom ibid.
Adjutants to attend the Town-Major till dismissed ibid.

CONTENTS.

Orderly Serjeants and Corporals to attend their Adjutants
Page 189
French Governors obliged to see the guards march off
ibid.
Dutch *Governors not, but often do* ibid.
Garrisons remote from the enemy, the orders delivered in the morning ibid.
In frontiers, not till the gates are shut ibid.
Upon the arrival of a person entitled to a guard, the eldest Regiment in garrison is to furnish it ibid.
ARTICLE V. *Directions for the mounting and dismounting of guards* 190
Corporals of guards to inspect into the things committed to the charge of their Sentries ibid.
How they are to proceed when any of those things are damaged or lost 191
ARTICLE VI. *Manner of relieving Sentries* 193
Directions how Sentries are to behave on their posts 194

CHAP XII.

Instructions to the Officers on guard till they are relieved.

ARTICLE I. *The Officers to keep their guards, and how they may give their men leave* 198
Officers on the port-guards to examine all strangers, and how they are to proceed with suspected persons ibid.
ARTICLE II. *Officers of the port-guards to send their reports to the Captain of the main-guard, with the time and manner of doing it* 199
Directions to the Captain of the main-guard, about entering of the several reports, and time of delivering them to the Governor ibid.
Reserve-Guard, only under the direction of the Governor 200
ARTICLE III. *Time the draw-bridges are kept up, and the barriers shut* 201
Port-

CONTENTS.

Port-guards to be under arms, when troops enter the town. Reasons for it — Page 201

Upon a fire, the port-guards to be under arms, and the gates shut — 202

Reason for it — ibid.

The same precaution when a riot happens near a port-guard — ibid.

Main-guard to quell those in the middle of the town — ibid.

In frontier garrisons the guards to be doubled on market-days. Reason for it — ibid.

ARTICLE IV *Port guards to send for the keys at sun-set, and the Drummers to beat the* Retreat *on the ramparts* — 203

The gates to be then closed, and the wickets only left open — ibid.

No Soldier to have port-liberty then — ibid.

The keys brought to the main-guard, and delivered to the Serjeants of the ports — ibid.

Port-guards, how they receive the keys, and ceremony of shutting the gates — ibid.

How guards are to be posted in the out-works during the night — 204

When they are numerous, a method proposed for the doing of it — ibid.

When the gates are locked, how the keys are to be returned — ibid.

Manner of delivering of the night-orders — 205

Captain of the main-guard to distribute the tickets for the rounds — ibid.

Additional visiting rounds, on what occasions appointed — ibid.

Manner of doing it, and by whom — ibid.

Time of beating the Tat-too, *and by whom* — 206

All Soldiers to retire to their quarters at Tat-too — 206

Burghers to carry lights with them, when they go out after Tat-too — 206

Penalty if they do not — ibid.

CONTENTS.

ARTICLE V. *Patroles to go every hour after* Tat-too, *and duty of the patroles* Page 206

Proper districts to be assigned for for the patroles of each guard 207

What places are most proper for the horse to patrole in ibid.

Serjeants of the patroles to make a report to their Officers ibid.

ARTICLE VI. *Ordinary and extraordinary rounds* 208

Town Major's round to go first, with the design of it ibid.

Manner of going and receiving of rounds ibid.

Reason for the said ceremonies, to shew the necessity of keeping them up 210

Town-Major to make a report to the Governor after his round 212

In France, *the Officers give the parole to the Town-Major* 213

Their reason for it ibid.

When the Rounds or Sentries discover any men from the ramparts, what they are to do 215

How the Officers, who command the guards, are to act on such notice ibid.

The design and duty of rounds 216

ARTICLE VII. *At* Reveillé *the keys to be carried to the main guard, and delivered to the Serjeants from the ports* ibid.

Patroles of horse to be sent with the keys 217

Precaution in letting the patroles of horse out of the gates ibid.

Design, manner and time of their patroling ibid.

When the patroles return, the gates to be opened, and the keys sent back 218

Port guards to remain under arms while the horse are patroling ibid.

CHAP.

CONTENTS.

CHAP. XIII.

Of detachments, visiting the Soldiers quarters and the hospital.

ARTICLE I. *Upon what occasion detachments are ordered* Page 219
A separate duty from the town-guards ibid.
Manner of appointing Officers and Serjeants for those commands ibid.
Every battalion to furnish an equal proportion ibid.
When Officers are detached, their ordinary complement of men 220
Contrary to the rules of war to send less than nineteen men ibid.
ARTICLE II. *Not to pass for a duty unless they march beyond the barrier* 221
How an Officer is to mount guard when he returns from a party ibid.
Less than an Officer and twenty-five men should not be detached but on necessity ibid.
All detachments to have passports from the Governor ibid.
Consequence if they have not, or consist of less than nineteen men 222
Reason, supposed, for this custom, and how far it should be adhered to ibid.
How an Officer on party is to act, when he sends a detachment from him 223
All parties to return at the time limited, unless prevented by the enemy ibid.
ARTICLE III. *Orderly Serjeants and Corporals, their duty* ibid.
The Soldiers divided into messes, and manner of doing it 224
Necessary to see their victuals dressed. Reason for it. Consequence if neglected ibid.
Directions about visiting the soldiers quarters 225

CONTENTS.

ARTICLE IV. *Directions about visiting the sick Soldiers. Incumbent on the Captains to assist their own men, and not let them remain too long in the hospital* Page 226
Where there is no hospital, an infirmary to be appointed 227

CHAP. XIV.

Command of the Governor. Compliment due to him, and the other Officers, from the Troops in Garrison.

ARTICLE I. *The Governor has the chief command of the Troops in his Garrison, though Officers of a superior rank should be ordered in with them* 229
Who the command falls to in the absence of the Governor, &c. ibid.
His power over the civil, determined by the constitution of the country 231
That over the military much more extensive ibid.
Officers and Soldiers not to lie out of the Garrison without his leave ibid.
General method of granting his leave of absence 232
His power not limited in this, consequences if it was ibid.
How Officers and Soldiers are to apply for leave of absence 233

ARTICLE II. *Soldiers who have leave must have passports signed by the Governor* ibid.
Those who go without to be taken up and tried as deserters 234
Officers on the port-guards to examine all Soldiers who enter the gates ibid.
Officers on party to examine all Soldiers they meet ibid.
Regiments cannot hold Courts Martial without leave from the Governor ibid.
Nor appear under arms but by the said leave, and mentioned in orders ibid.
Reason, supposed, for this custom 235
ARTICLE III. *On an alarm, all the Troops to repair to their alarm posts* ibid.

How

CONTENTS.

How the Colours are to be conducted there Page 236
On what occasions the Troops may be ordered to their alarm-posts ibid.
When assembled, none can dismiss them but the Governor ibid.

ARTICLE IV. *Compliment paid the Governor by the Guards* 237
Compliment due to General Officers and others, in camp and garrison 237
Compliment paid to a Marshal of France *in the* French *garrisons* 239
His authority superior to all Governors, even in their own garrisons ibid.
A copy of the late Duke of Marlborough's *commission of Captain-General* 240

ARTICLE V. *Governors of citadels, &c. to send for the word from the Governor of the town* 242
Governor of a citadel, his power over his own garrison ibid.
Not to suffer above one third of his garrison to go out at a time ibid.
Governors of towns and citadels in France, *their command seperate, &c.* 243
Common to find it so there, and reason alledged for it ibid.

ARTICLE VI. *Town-Major and his Adjutants to view the fortifications, &c. frequently* ibid.
Inhabitants to give in the names of those who lodge with them, &c. ibid.
A table for the several duties in a garrison 244

CHAP. XV.

Duty of the Infantry in camp.

Of guards ordinary.

ARTICLE I. *What is generally meant by camp duty* 245
Guards ordinary, what they are, and numbers they generally consist of ibid.

CONTENTS.

ARTICLE II. *Method of mounting and dismounting of the quarter-guard, duty and design of it* Page 246

ARTICLE III. *Directions for the parading the men of the other guards ordinary* 250

ARTICLE IV. *How the guards are formed on the grand parade* 252

ARTICLE V. *Picquet-guard, the design, and time it continues on duty* 254

Time of drawing out, with the Captain's directions to the men 256

Men of the picquet not to be changed, and reasons why 257

How far the picquet is to march when it passes for a duty ibid.

General Officers of the day have the immediate command of it ibid.

Order for the marching of the picquet, to be sent to the Lieutenant-General of the day 258

According to the number of Regiments, Field-Officers are appointed for the picquet ibid.

Infantry divided into wings for the ease of appointing them ibid.

When the army is large, each wing furnishes its own Field-Officers ibid.

Manner of appointing them to their several commands 259

A plan of a body of foot formed into two lines, divided into Wings and Brigades, to shew how the Field-Officers are appointed for the picquet

Colonels of the picquet go the grand round, the other Field-Officers go theirs, as they shall direct, and when to make their reports 260

General Officers of the day always received as grand rounds ibid.

None but the Lieutenant-General of the day can draw out the picquet by night ibid.

Upon an alarm, a method proposed of joining the picquet 261

ARTICLE VI. *In what the* German *and* French *picquets differ from ours* 262

CONTENTS.

Method of the Foot Picquet of the Germans, *with reflections on it.* Page 262
Method of their Horse Picquet, preferable to that of the foot 265
Reason why, and how it may be improved 267
When followed, the necessary precautions which should be taken ibid.
Method of the foot picquet of the French 269

CHAP. XVI.

Guards ordinary of the cavalry, and guards extraordinary of the foot.

ARTICLE I. *Guards ordinary, standard-guard, number they consist of* 271
Remarks on their mounting in boots 272
ARTICLE II. *Grand-guard, number unfixed, divided into Captains commands. How posted* ibid.
Design and duty of the grand-guard 273
Manner of relieving it 274
Day and night posts of the grand-guard, and reason why ibid.
To whom the posting of the grand-guards belong 275
In proportion to the number of the grand-guard, Field-Officers are appointed 276
Directions to the Officer who commands it, when the enemy approaches ibid.
To whom he is to make his report when relieved 277
ARTICLE III. *Guards extraordinary of the foot, on what occasions ordered* ibid.
What is meant by out-posts, and reason for them ibid.
Usual time of relieving out-posts ibid.
Adjutants to see the men who mount, supplied with bread and pay 278
Out-posts, necessary precautions for their security ibid.
Their orderly men to attend the Major of Brigade of the day. Reason why ibid.

CONTENTS.

Out-posts, how they are to receive the Generals who visit them Page 279

Those near the camp to have the parole which is given to the army ibid.

Those at a distance to have one of their own, and by whom sent ibid.

Necessary conduct of an Officer who commands an out-post ibid.

How far he is to maintain his post, must depend on his orders and situation 280

Custom of war in these cases 281

ARTICLE IV. *Foraging parties, design and duty of them* 281

ARTICLE V. *Convoys, or escorts, on what occasions ordered* 283

ARTICLE VI. *Expeditions, nature and design of them* ibid.

CHAP. XVII.

General rules for encamping an army, &c.

ARTICLE I. *Proportions to be observed, in encamping a Regiment of dragoons, of six Troops, forming three Squadrons; with the Light Troop* 285

The dimensions of the Captains and Subalterns tents 289

ARTICLE II. *Common proportions of a trooper's tent, with directions for encamping a Regiment of horse, or dragoons, of nine Troops, and a plan of the same* 291

Duty of the Quarter-Master, and camp-colour-men 295

ARTICLE III. *The several beatings used the day the army marches, and what is to be done at each* 296

ARTICLE IV. *Duty of the Quarter-Master-General, with the method of drawing out an encampment, or line of battle* 297

CONTENTS.

CHAP. XVIII.

Duty of the troops at a siege.

ARTICLE I. *The foot employed in making the trenches, but paid for the same.* Page 304
The prices paid during the late war ibid.
Time the men work in the trenches 305
Fascines, gabions, &c. made by the foot, with a description and proportion of them, and method of carrying them to the magazine 307
ARTICLE II. *Directions to the covering-party and workmen, at the opening of the trenches* 310
Manner of making trenches, with the common proportion of the several parts 313
When the workmen leave off, the covering party takes possession of the trenches 315
Directions to the Officers who command the workmen, when a Sortie *is made* 316
ARTICLE III. *Guard of the trenches, done by entire Battalions, and commanded by General Officers* 317
Manner of relieving the guard of the trenches, with directions to those who mount, and the necessary precautions to be taken 319
Practice of the French *Battalions when they mount the trenches* 322
ARTICLE IV. *A guard of horse commanded to each attack, but remain at the Queuë of the trenches* 324
When a sally is made, the horse are to assist the foot in repulsing the enemy 325
Method of the French *Governors, by making small Sorties to retard the progress of the besiegers* 326
When any little attack is to be made, it is to be done by the guard of the trenches, and the Battalions on duty give an equal number of men for that service 327
ARTICLE V. *General rules for a considerable attack, with the disposition of the troops who make it* 329

CHAP.

CONTENTS.

CHAP. XIX.

Manner of receiving and diftributing the daily orders, with the general detail of duty, according to the method in Flanders.

ARTICLE I. *General Officers of horfe and foot, feparate commands and duty* Page 334
Diftinct General Officers of the day for the horfe and foot ibid.
Number of each, and time they remain on that duty ibid.
Generals of the Day command the picquet, and vifit the grand-guard and out-pofts 335
Majors of Brigade roll for the day, and duty of him who is of the day ibid.
Duty of the Adjutant-General 336
ARTICLE II. *How the orders are received and diftributed* 337
ARTICLE III. *How the duty of General Officers and others is regulated; as alfo that of entire Battalions* 340
Plan of a rofter, or table, by which the duty of Field Officers, Captains, Subalterns, entire Battalions and Squadrons is regulated, with an explanation of the rofter 345
Table of proportion for the detaching of private men, with an explanation of the table 348

CHAP. XX.

Manual Exercife and Evolutions of the Cavalry explained.

ARTICLE I. *Proper arms of a trooper, and manner of pacing them* 349
Directions for the forming and drawing up of Squadrons 350
The feveral diftances of ranks and files 351
How the Officers poft themfelves at the head of the Squadrons ibid.

How

CONTENTS.

How the Standards are brought to the Regiment and returned Page 352

How the Squadrons are to be told off, or divided for the exercise 354

General rules for wheeling 355

ARTICLE II. *Words of command for the Manual Exercise, and Evolutions of the Horse, with the explanation, calculated for a Regiment of 3 Squadrons* 357

ARTICLE III. *Words of command for the Exercise of a Regiment of two Squadrons* 368

ARTICLE IV. *Evolutions of the Horse, with an explanation* 370

CHAP. XXI.

Exercise of the Dragoons explained

The necessary alterations in the points where they differ from the horse and foot 379

CHAP. XXII.

Part of the Exercise for the Troops of Light Dragoons 389
Evolutions of the Light Dragoons 393

CHAP. XXIII.

ARTICLE I. *Rules for the reception of a General Officer, who comes to review the Regiment, or see the Exercise performed* 396

A

CONTENTS

That the Standards are brought to the Regiment and received .... Page 355

How the Standards are to be rolled, unfolding them for an engagement 356

General rules for unfolding 357

Article LIV. What is necessary for the Mayor at Exercise and Exercises of the troop, with the explanation and remarks for a thorough guidance 357

Section 1III. Words of Command for the Exercise of a Regiment of twelve squadrons 368

Article LV. Description of the harness, and its explanation 369

CHAP. XXI.

Exercise of the Dragoons explained

How dragoons are to be in the ranks, were they alight from the horse and foot 370

CHAP. XXII.

Article. Exercise for the Use of Light Dragoons. 386
Evolutions of the Light Dragoons. 397

CHAP. XXIII.

Article LVI. Rules for the treatment of General Officers, who come to review the Regiment, or the H. Regiment. 390

A TREATISE

OF

Military Discipline.

CHAP. I.

Containing directions for the forming of Battalions, posting of Officers, &c.

ARTICLE I.

WHEN the battalion is ordered to draw out to exercise, or upon any other occasion, the soldiers are to repair, at the time appointed, to their Captain's quarters, or the place ordered for that purpose, completely armed and accoutered. The subalterns, and non-commission Officers, are to be there at the same time.

As soon as the men are paraded, the Lieutenant, or in his absence the Ensign, must order them to stand to their arms, and form them into a rank entire, or three, or six deep; after which he is to view their arms, ammunition, clothes and accoutrements, and to see if they are clean and dressed in a soldier-like manner, in order to make a report of each particular to his Captain, when he comes to march the company to the general parade.

As there is not any one thing which contributes more to the appearance of a regiment than the sizing of the men, great care should be taken in the doing of it, before they march to the general parade. For which end there should be a size-roll made for each company, that the men may know what rank they are to form in, which will save the Officers both the time and trouble of doing it; for, by casting their eye over the size-roll, they can immediately see if the men are drawn up according to that, or not.

The sizing of men, is the placing, as near as possible, those of an equal height into each rank.

In forming six deep, the tallest men must be placed in the front and fourth ranks. The reason for it is, that when the rear half-files are doubled up, the men of the fourth rank may size with those of the first. The same rule must be observed in sizing the men of the second and fifth ranks; as also those of the third and sixth, without which precaution, the battalion will appear to a great disadvantage.

The easiest and most exact method for the doing of this, is to draw up the companies at the Captain's quarters three deep, placing the tallest men in the front and rear-ranks, and the lowest in the center.

In the forming of four deep, (which is seldom done but when a battalion is very weak) the same regard must be had, that the ranks may appear equally sized when they double.

The companies being formed and sized, the men's arms, &c. inspected according to the above directions, they are to be told off into two or four divisions, and marched, with shouldered arms, by their respective Officers, to the general parade, or rendezvous of the regiment. If they march in two divisions, the Captain leads the first; the Ensign the second; and the Lieutenant brings up the rear. If in four divisions, the Ensign leads the third.

Note.

Chap. I.   *Military Difcipline.*   3

Note. A company is here fuppofed to confift of no more than feventy private men, with two fubalterns, three ferjeants, three corporals, and two drummers, which is the fixed eftablifhment for Britifh infantry: And although the number has been augmented to a hundred, fince the commencement of the prefent war, together with an additional Lieutenant, ferjeant, and corporal, yet it is only on account of that extraordinary demand for men, which a war muft always create, and will hardly fubfift any longer, than while it lafts.

The ferjeants are to march on the flanks, and to fee that they carry their arms well, and keep their ranks ftraight.

The eldeft ferjeant takes poft on the right of the front rank of the firft divifion; the fecond ferjeant on the left of the front rank of the laft divifion; and the youngeft, on the right of the front rank of the divifion which the Enfign leads. The drummers are to fall in between the firft and fecond ranks of the firft divifion; but when they march fix deep, they are then to fall in between the third and fourth. In this order the companies are to march whenever they march with fhouldered arms, except at reviews; the difference in which fhall be fhewn in its proper place.

The pioneer, with his firelock refted on his left arm, marches at about twelve paces before the Captain. The ranks, if at open order, are twelve feet, or fix paces afunder; at clofe order, they are at two feet, or one pace afunder.

When the companies come to the parade, or place where they are to form into battalion, they are to draw up according to feniority, thus: The Colonel's company on the right, the Lieutenant-Colonel's on the left, the Major's on the left of the Colonel's, the eldeft Captain's on the right of the Lieutenant-Colonel's, and fo on from right to left, till the youngeft comes in the center, leaving an interval of two paces, or four feet, between one company and another.

The grenadier-company draws up on the right of all,

at ten paces from the right of the Colonel's company. See the annexed plan.

As the companies draw up, the ſubalterns are to move up to the front, the Lieutenants placing themſelves on the right of their Captains, and the Enſigns on the left, towards the flanks of their companies. As ſoon as the Officers come up to the front, they are to face to the right-about, being at the diſtance of eight feet, or four paces from the front rank ; and when they have ſeen, that the half files cover well, and that the ranks are ſtraight, they are all to come to the left-about together, the Captain firſt giving notice to his ſubalterns. The ſerjeants are to form themſelves in the rear of the rear-rank, at the diſtance of four paces from the men, ordering their halberts. The drummers are to march about twelve paces in the front, where they face to the right-about, beating the *troop*, from the time they quit their reſpective companies, till the Officers are at their poſts.

The pioneer advances two paces beyond the drummers, and then faces to the right-about alſo.

The companies are to leave an interval of two paces between them, when they draw up in battalion, and to march in with their files complete, ordering their odd men, if they have any, to fall in the rear of the rear-rank, where the Serjeant-Major muſt immediately join them together, form them into files, and draw them up in the intervals between the companies, which they may do in an inſtant, by beginning on the right, and joining thoſe of two or three companies, or as many as will make a file, and then placing them in the firſt interval ; and ſo on till they are all formed.

If this method is duly obſerved, (which, from its being ſo eaſy and plain, can admit of no difficulty) it will effectually anſwer the end propoſed, that of ſaving both the trouble and time, which any other will of courſe take up ; a fault which ought to be avoided, that the ſoldiers may not be kept too long under arms, before they proceed to the exereiſe, in order to have it well performed.

Plan of a Battalion of Foot, drawn up according to Seniority of Companies; which are supposed to have just formed in Battalion, having their own Officers in their Front, and their Serjeants in their Rear; with their Files close, and their Ranks at the proper distance of 4½ feet, or 6 paces each. A Man is allowed 21 Inches to stand upon, and the Companies consist of 73 Rank and File.

Explanation

A. the Rank of Officers.
B. Front Rank of the Regiment.
C. Center Rank.
D. Rear Rank.
E. Rank of Serjeants.
F. Rank of Drummers.
G. Pioneers &c.

Note. The Drums, Fifes and Adjutant having no fixed Post assigned them, till the Battalion is completely formed, I shall therefore omit in this Plan.

Company of Grenadiers.
Colonels Company.
Majors Company.
2.d Captains Company.
4.th Captains Company.
5.th Captains Company.
3.d Captains Company.
1.st Captains Company.
Lieutenant Colonels Company.

Scale of Paces
10, 20, 30, 40, 50, 60, 70, 80, 90, 100, 110, 120, 130, 140, 150, 160, 170, 180, 190, 200, 210

220 Paces — or 440 feet

Chap. I. *Military Discipline.*

formed. Besides, the old method of completing the files of the whole battalion to the right and left, has, in my opinion, more the air of militia than soldiers.

The intervals which remain between the companies, after the odd men are formed, are to be closed by facing the companies to the right, which is to be done by word of command from the Adjutant.

The company of grenadiers is always to draw up on the right of the Colonel's, and to leave an interval of ten paces between them.

When the battalion draws up six deep, the grenadiers are to do the same: And when the battalion is drawn up four deep, for the punishing of the soldiers by making them run the gantlope, the grenadiers are to do so too; but if the battalion does it on account of its being weak, then the grenadiers generally draw up three deep.

ARTICLE II.

As soon as the odd men are formed, and the intervals closed, the Major is to order the Officers to take their posts in battalion; which is done by seniority of commission, thus: The eldest Captain is to place himself on the right of the battalion, the second Captain on the left, the third Captain on the left of the eldest, the fourth on the right of the second Captain, and so on, till the youngest Ensign comes in the center.

At the word of command for it, they are to face briskly (at two motions) to the right or left, according to the situation of their respective posts, recovering their espontons at the same time.

The serjeants, at the same time, recover their halberts, facing to the right and left outwards.

All the pioneers face to the left, as do also, one half of the drummers; the other half of the drummers face to the right, excepting the two drummers, who are to be orderly, and they face to the right-about.

At the word of command, *march!* from the Major, the Officers step off with their left feet, those who are to go to the left, marching on the outside. When they come

come to their respective posts, they are to stand fast, those upon the left of the colours remaining faced to the left, and those upon the right of the colours, to the right; six serjeants are to move up in two half files to the right flank of the battalion, and six to the left, dressing with the three ranks.

The remaining serjeants are to divide their ground equally in the rear, which circumstance must be carefully attended to by the Serjeant-Major.

The eldest grenadier serjeant, at this time, takes post upon the right of the front rank of his company; the next eldest, upon the left of the front rank, and the youngest remains behind the center of the rear rank.

The pioneers, having joined the pioneer of the Colonel's company, march till they are clear of the right of the grenadiers: then facing together to the right, they march, till they come to dress with the front rank.

The drummers of the right march on, till they join those of the Colonel's company; then having marched clear of the right of that company, they face to the right, and march, till they come in a line with the front rank.

The drummers of the left, having joined before the Lieutenant's Colonel's company, proceed to the left, till they are clear of the serjeants, then facing to the left, march till they come even with the front rank.

The two orderly drummers march, and post themselves upon the right hand of the Major.

The grenadier drummers march to the right of their own company.

When the Major gives the word of command, *Halt!* all the Officers, serjeants, pioneers, and drummers, face to the front, in doing which, the Officers and serjeants are to order their arms.

The Officers must be careful to dress themselves in a straight line, at the distance of four paces from the front rank, dividing the ground equally between them, and every one covering a file exactly.

The Officers of grenadiers, with the serjeants and drummers, remain with their own company, and dress
with

Chap. I. *Military Discipline.* 7

with those of the battalion. The Captain posts himself in the center, the eldest Lieutenant on the right, and the second Lieutenant on the left of the company.

When the battalion is drawn up, the Colonel's post is in the center, and about six paces before the Ensigns with the colours. The Lieutenant-Colonel's post is a little to the left of the Colonel's, and about two paces from the rank of Officers, that the Colonel may be four paces advanced before him.

When there is no General, or superior Officer present to see the exercise performed, the Colonel does not take his post at the head of the battalion, but remains in the front, without taking his esponton in his hand, to give the Major the necessary orders about the exercise, &c. in this case the Lieutenant-Colonel is to post himself at the head of the battalion with his esponton in his hand; but if the Colonel is absent, he then remains in the front in the same manner as before-mentioned for the Colonel, the sole command then devolving on the Lieutenant-Colonel.

Whenever the regiment is drawn out, no Officer must be absent without leave from the commanding Officer; but each stand at the post assigned him in the rules of excercise, with his arms in his hand, expecting such orders as he may receive, either to exercise the whole battalion, a part of it, or a single company, as the commanding Officer shall direct; which he is always to perform with his esponton in his hand.

ARTICLE III.

The Officers having taken their posts in the front of the battalion, and the ranks and files being dressed, the colours are then to be sent for; which is usually performed in the following manner:

The Major is to order one of the Grenadier drummers to beat the drummer's call; upon which the Ensigns, who are to carry the colours, are to repair to the head of the company of grenadiers. The Captain, his two Lieutenants, and the company of grenadiers, are to march with the Enſigns, and guard the colours to the regiment. They march with their firelocks reſted on their left arms without fixing their bayonets, till they receive the colours. The words of command to the grenadiers (and which are given by the Captain) are as follows:

I. *Poiſe your firelocks!* - - - - 2 ⎫ number of
II. *Reſt your firelocks on your left arms!* 1 ⎬ motions.

When this is done, the Captain places himſelf two paces before the Lieutenants, and marches to the Colonel's quarters, or place where the colours are lodged; the drummers, marching in the rear of the Enſigns, beat the *troop*, and the company of grenadiers, in two diviſions, marches immediately after the drummers; one ſerjeant marches on the right flank of the front rank, of the firſt diviſion; one on the left flank; and the third on the right flank of the front rank of the ſecond diviſion, with their halberts advanced.

Note, In ſending for the colours, the compliment then paid by the battalion, is that of ſhouldered arms; but when the colours are brought to the regiment, they are received with reſted arms, and the drummers beat a *march*. The ſame compliment is paid to the colours when they are ſent from the battalion; but when the Enſigns go for the colours, and return without them, the battalion is only to be ſhouldered.

As ſoon as the Captain comes to the place where the colours are lodged, he is to draw up his company three deep, and then order the grenadiers to fix their bayonets, as follows:

I. *Poiſe your firelocks!* - - - - 2 ⎫
II. *Reſt on your arms!* - - - - 3 ⎪
III. *Draw your bayonets!* - - - - 2 ⎬ number of
IV. *Fix your bayonets!* - - - - 3 ⎪ motions.
V. *Poiſe your bayonets!* - - - - 3 ⎪
VI. *Shoulder!* - - - - - - - 2 ⎭

When

When the Enſigns receive the colours, the captain gives the word of command.

VII. *Preſent your arms!* 3 motions.

Upon which the grenadiers preſent their arms, and the drummers beat a *point of war*; after which the Captain gives the following command:

VIII. *Recover your arms!* - - - 1 ⎫ number of
IX. *Reſt your bayonets on the left arm!* 2 ⎬ motions.
X. *To the right* (or *left*) *wheel! march!*

This being performed, the Captain marches back to the battalion in the ſame order that he came from it, the Enſigns carrying the colours advanced, and letting them fly.

As ſoon as he approaches the left flank of the battalion, the major gives the word of command, *Face to the left!* upon which the battalion faces to the left, ſtill remaining with reſted arms, and the drummers beating a *march*. When he has marched paſt the left flank, he is to make two wheels to the left, ſo as to come up to the front of the battalion, the ſecond diviſion of grenadiers at this time moving up to dreſs with the firſt. In this poſition he is to remain till the Major gives the word to the battalion. *To the right as you were!* at which he, together with the Lieutenants, Enſigns, drummers, and the three ranks of grenadiers, faces to the right, together with the battalion, and immediately after marches off, followed by the Lieutenants, and behind them, the Enſigns with the colours before the line of Officers. The drummers march between the Officers, and the front rank of the battalion; the front rank of the grenadiers between the front and center rank of the battalion; the center rank of grenadiers between the center and rear; and the rear rank of grenadiers along the rear of the battalion. When the Enſigns come to their poſts in the center, they are to fall in, the Colonel's colours

colours taking the right, the Lieutenant-Colonel's the left, and the Major's (if there are three colours) the center; when the Ensigns have taken their posts, the Major orders the battalion to shoulder. The grenadiers keep marching on, till they come to their former ground, upon the right of the battalion, where the Captain is to order them to halt, and immediately proceed to unfix the bayonets, by the following words of command:

I. *Poise your bayonets!* - - - - 2	⎫	
II. *Rest on your arms!* - - - - 3	⎪	
III. *Unfix your bayonets!* - - - 2	⎬	number of motions.
IV. *Return your bayonets!* - - - 2	⎪	
V. *Poise your firelocks!* - - - 3	⎪	
VI. *Shoulder your firelocks!* - - 2	⎭	

When the parade is at any considerable distance from the place were the colours are lodged, it is usual for the Field-Officers companies to assemble there, and carry them with them to the place of exercise, that no time may be lost in the sending for them. In this case, the Captain-Lieutenant marches at the head of the first division, and the Ensign, who carries the Colonel's colours, at the head of the second division; the soldiers having their arms shouldered, and the drummers (who fall in between the front and center ranks of the first division) beating a *march*. The Lieutenant-Colonel's and Major's companies, if there are three colours, are to observe the same method in carrying their colours with them; and if the proper Officers belonging to those companies are absent, others must be ordered to them for that purpose.

When the colours are brought in this manner, they are to remain with the companies till the Officers are ordered to take their posts at the head of the battalion, at which word of command, the Ensigns are to march with them to the center.

As soon as the colours are to be sent back, (or lodged, according to the military phrase) the drummer's call is to be be beat at the head of the grenadiers; on which the Enfigns with the colours are to repair thither immediately, and draw up as before. The Captain of grenadiers is then to order his company to fix their bayonets, and reſt them on their left arms: And as ſoon as the Major has ordered the battalion to preſent their arms, he is to march back the colours to the place, where they are to be lodged, the drummers beating the *troop* as before. The Enſigns are to carry the colours back in the ſame manner they brought them, that is, advanced and flying; and as ſoon as they arrive at the place, and the company is drawn up, they are to furl the colours and lodge them. When this is done, the Captain is to order the grenadiers to unfix, and return their bayonets, and reſt their firelocks on their left arms; after which he is to march back in the ſame manner he carried the colours to the battalion, unleſs he is ordered to diſmiſs his men as the colours are lodged; in which caſe, when the bayonets are returned, inſtead of poiſing, he is to order them to reſt their firelocks, then club, and diſmiſs them with the ruff of a drum.

ARTICLE IV.

The ſeveral diſtances of ranks.

In the drawing up of a battalion for exerciſe, or a review, the ranks are to be at ſix ordinary paces diſtance from one another.

When they are to fire, either by ranks, platoons, the whole battalion, or in the ſquare, the ranks are to be moved up to one pace diſtance.

In all wheelings, either by diviſion, or the whole battalion, the ranks are to be cloſed forward to cloſe order, which is to one pace diſtance.

Diſtance

Distances of files.

When a regiment is to exercise, or to be reviewed, the files are to be opened, the distance of which between one another, is one pace, or the length of an out-stretched arm; but that this may appear more plain, as soon as the files are opened, and the men faced to their proper front, order those of the front rank to stretch out their right arms to the right, and if they can touch the left shoulder of their right-hand men, they have then their true distance; the doing of which now and then, will give them a just notion of their proper distance. As the men of the rear-ranks are to be governed by, and dress with those in the front, who are called their file-leaders, there is no occasion for their doing it.

In firing, marching or wheeling, the files must be so close, that the men touch one another with their shoulders. These are all the distances of ranks and files which are necessary to be known for the performing of every part of the service now practised.

Rules for the opening of files.

In the opening and closing of files, or marching all together to the right or left to change their ground, they should begin with the feet next the front.

In facing to the right, they are to do it, by raising the right foot, and planting it four inches behind the left heel, at the first motion; and by facing to the right upon both heels, at the second.

In facing to the left, they are to bring the right foot forwards, placing the heel against the inside of the left toe at the first motion; and by facing to the left upon both heels, at the second.

As soon as the men of the first file on the left of the battalion lift up their right feet a second time, the men of the second file are to lift up their right feet; and when those of the second file lift up their right feet a second time,

time, the men of the third file are to begin with their feet. All the other files are to follow the same directions till the whole have opened.

In the doing of this, they are to take but short steps, and to move on very slow, but with an equal pace, lifting up and setting down of their right and left feet with one another, thus: All who are in motion, must lift up and set down their right feet together, and do the same with their left.

The bringing of a battalion to such exactness as to perform it in due time, will, I am afraid, appear so difficult, that it will deter a great many from attempting it; but let those, who are of this opinion, only try, and they will find it much easier in the execution than they imagined.

The common objection against it, is, that it looks too much like dancing, and makes the men appear with too stiff an air. I own it may have this effect in the beginning; but a little time and practice will bring the men to perform it in so easy and genteel a manner that the objection will vanish. A great many reasons may be brought to support this argument, were there an occasion for it, such as the bringing of the men to walk with a bolder air, giving them a freer use of their limbs, and a notion of time; which, in my opinion, are sufficient to silence those who oppose it; and therefore I shall not trouble the reader any further, but proceed to the point in hand.

When the battalion is six deep, the men of the five rear-ranks must keep in a straight line with their file-leaders, and neither close nor open their ranks in marching; and all carry their arms high and firm on their shoulders, looking up and keeping their bodies straight.

The Officers are to carry their espontons downwards, and the Ensigns to advance the colours.

No Officer is to move till the file opposite to him does; and when the men of that file step forward with their right feet, he is to do the same with his, keeping an equal pace and in a direct line with them; by the observance

servance of which rule, the Officers will, when they halt, have their proper distance between them, provided it was equally divided before the files were opened, which will save them the trouble of moving afterwards.

The serjeants in the rear must observe the same rule; only they must carry their halberts recovered.

As soon as the second file on the right of the battalion, which is the last to open, steps forward, the Major, or commanding Officer, is to proceed to the following word of command.

HALT!

On the giving of this word of command, the Officers and soldiers are to face briskly to the right at two motions, placing the right foot about four inches behind the left heel at the first, and facing to their proper front at the second, in the manner above described. As soon as they are faced, they are to dress their ranks and files.

ARTICLE IV.

After the Ensigns with the colours have taken their posts, and the escort of grenadiers have returned their bayonets, and shouldered, in the manner above-directed in Article III, the Major will form the battalion six deep, by making the even, or rear half-files to face to the right, and march eighteen paces to the rear. The half-files of serjeants recover their halberts at the same time, then step off with the right feet, when the half-files do, which they cover; coming about also when they do, and ordering their halberts. After the Major has made them *halt*, the Adjutant is to assist him, in going from the right to the left of the battalion, dressing most exactly every rank, and every file, and seeing that the Officers, as well as the serjeants, be always in a line with the files.

The battalion thus formed, and dressed, the Colonel will take his post six paces before the center of the line

line of Officers; the Lieutenant-Colonel's post is two paces before the line of Officers, and a little to the left of the Colonel, (when he is present) or just before the center, when he is absent.

The Major will then order the battalion, 1st, to poise their firelocks; 2dly, to rest upon their arms; 3dly, to draw their bayonets; 4thly, to fix their bayonets; 5thly, to poise their firelocks: and, 6thly, to shoulder.

This being done, he will order the battalion to present their arms. All the drummers beat the *march*, as soon as the men come down to their rest; at which time, the Major, raising his sword, and dropping the point, gives the signal for all the Officers to salute together, and the Ensigns to drop their colours; the whole having pulled off their hats together are to remain so, till the Major raises the point of his sword again, upon which the Officers are all to put on their hats at the same time.

This is a part of the exercise, and no compliment to any general Officer, who may be reviewing the regiment, the reception due to whom, must take place before it, and will be particularly explained hereafter.

The Major is then to order the battalion, 1st, to poise their bayonets; 2dly, to rest on their arms; 3dly, to unfix their bayonets; 4thly, to return their bayonets; 5thly, to poise their firelocks; and, 6thly, to shoulder; which being done, he will proceed to the Manual Exercise.

ARTICLE VI.

When the regiment draws out, the Major and Adjutant should be always on horseback, it being impossible for them to perform their duty on foot, in the manner it ought to be. They are likewise to exercise the regiment on horseback; but no other Officer is to do it mounted, unless such, who, in their absence, are appointed to do their duty.

When

When the commanding Officer orders any of the other Officers to exercise the battalion, or a part of it, they are to do it on foot with their half-pikes or espontons in their hands. This should be frequently done for the instruction of the Officers, that, upon the absence of the Major and Adjutant, they may have a number sufficiently qualified to act in those posts.

Whoever exercises a battalion, or any number of men, should place himself opposite to the center, and at such a distance, that he may take in the whole at one view; but if that distance is too far off for his voice, he may place himself nearer, though still opposite to the center.

As the performing of the exercise well, depends a great deal on the giving of the words of command; those who exercise, should deliver the words clear and distinct, that the men may not mistake one command for another.

Whoever would attain to any perfection in it, must study the compass of his voice, that he may not overstrain it, lest it cause an immediate hoarseness; as also the laying of the emphasis in the right place, and where to make the proper stops, or pauses, when the command is too long to be pronounced at one breath. Besides, stops or pauses, when judiciously made, are of great service to the men, in giving them time to think on the word of command before it is fully delivered; and, consequently, preparing them for the performing of it with life, vigour, and exactness. But as the coming to this knowledge will prove tedious to young Officers, without some further assistance than that of mentioning it, I have placed comma's where the stops or pauses should be made, which, if observed, will aid and assist an indifferent voice, and give beauty and force to a good one; and enable every one to give the words of command with more ease to themselves, and clearness to the men. The rest must depend on practice, precept alone not being sufficient to arrive at perfection in any art.

Directions

Directions for the position of a soldier under arms.

Every soldier will give the greatest attention to the words of command, remaining perfectly silent and steady; not making the least motion with head, body, feet or hands, but such as shall be ordered; the heels, at this time, are to be in a line, not more than four inches asunder; the toes turned out; the shoulders square to the front, and kept back; the breast pressed forward; the belly drawn in, but without bending; the right hand hanging down on the right side, the back of the hand to the front; the firelock carried on the left shoulder, the barrel outwards, the butt in the left hand, two fingers being under it, the middle finger just upon the turn or swell, the fore finger and thumb above the swell, the piece almost upright, the butt flat against the outside of the hip-bone, the lock a little turned up, the guard being just below the left breast, the piece pressed to the body, with the ball of the thumb, the head turned a little to the right, except the right hand man, who looks full to the Major, or exercising Officer.—Great care must be taken, not to begin a motion, till the word of command, or signal on the drum, be ended, and then to be very exact in counting a second of time, or 1, 2, betwixt each motion. The Major, or exercising Officer, is to take the space of two seconds betwixt the end of each motion, and his giving the word of command, or signal, for ano ther; and this the men are likewise to observe, when they exercise by one word of command only.

CHAP

CHAP. II.

The exercise of a battalion of foot, with an explanation.

1st caution. FFICERS *take care!* or a roll upon the drum.

Recover your arms! or a flam on the drum. 2 motions.

On this, all the Officers, and serjeants, recover their arms in two motions.

To the right about! or a flam. 3 motions.

The Officers go to the right-about in 3 motions, except the Colonel, or commanding Officer of the battalion; the serjeants in the rear, face to the right, and left outwards.

March! or a troop. 43 motions.

The Officers march together, keeping the same line, and stepping with the same feet, until they are nine paces beyond the rear rank, except the Colonel, or commanding Officer of the battalion, who is to advance forward, opposite to the center of the battalion, two paces beyond the Major. The serjeants march at the same time, that the Officers do; those on the flanks, directly forward, followed by those from the rear until they have passed the line, where the Major stands, about fifteen paces; then face to the right and left inwards, and march unto each other, dividing the length of the front equally amongst them, but a serjeant must remain upon the right and left flank of each rank; the youngest serjeant

jeant of the grenadier company having faced to the right, marches to the front, dressing in a line with the rank of serjeants, but keeping opposite to the center of his own company. The second grenadier serjeant, having faced to the right-about, at the second word of command, marches to the rear, and moves up to the right flank of the rear rank. The pioneers do not move. The two divisions of drummers on the right and left of the battalion, and the musick, move directly forwards, beating the *troop*, till they have passed the line where the Major stands, about six paces: then, facing to the right and left inwards, move on till they are directly behind the Major, the musick being in the center. The Grenadier drummers march forward at the same time, and form on the same line, with the other drummers, but opposite to their own company.

Halt! or a flam.   3 motions.

Upon which all the drummers cease beating: the Officers face to the left-about in 3 motions: the serjeants, who had halted upon the feet next to the battalion, face upon the feet they halted on: the drummers, and musick, face to the battalion in like manner.

Order your arms! or a flam.   2 motions.

The Officers and serjeants order their arms in two motions. The Ensigns, who carry the colours, are allowed to order, or plant them, during the Manual Exercise; but, at all other times, are to carry them advanced: the grenadier Officers are likewise permitted to order their fusees.

The Manual Exercise.

Take care to perform the Manual Exercise! or a ruffle on the drum.

I. *Rest your firelocks!* 3 motions.

Join your right hand, by seizing the firelock behind the lock, at the same time turning it with the left hand, so as to bring the lock upwards, the piece being almost right up and down. Tell 1, 2, and come to the recover, by raising the firelock from your shoulder with your right hand, at the same time turning the barrel inwards, and seizing it with the left hand just above the feather-spring, that elbow close, the hammer about the height of, and just before the right breast, the right thumb upon the cock. Tell 1, 2, and turning upon both heels, so as the right toe may point to the right, and the left a little to the front, come down to the rest, the cock at the waist-belt, the left arm close to, and directly across the stomach, the right elbow turned towards the front, the butt of the firelock against the middle of the right thigh, the muzzle pointing very little forwards, the stock in the left hand, the right thumb under the cock, the knuckle of the fore-finger under the guard.

II. *Order your firelocks!* 3 motions.

Quitting the firelock with the left hand, seize it immediately as high as your shoulder with a brisk motion, at the same time bringing the firelock perpendicular, the butt being near to, and on the out-side of the right thigh. Tell 1, 2, and let go the right hand, sinking the firelock with the left, at the same time seize your firelock briskly with the right hand near the muzzle, the hand at the height of the chin, the thumb upwards. Tell 1, 2, and, turning your toes to their proper front, quit the firelock with the left hand, bringing down

down the butt-end to the ground, even with your toe, at the out-side of your right foot, the firelock upright, your right arm hanging, from the hand to the elbow, by the side of the firelock; the left hand hanging by the left side; and both shoulders square to the front,

III. *Ground your firelocks!* 4 motions.

Lift up your right foot, and, making a half face to the right, place it against the flat side of the butt, at the same time turning the barrel of the firelock towards your body. Tell 1, 2, and step directly forward with your left foot, quit the firelock with your right hand, and seize it immediately at the middle of the barrel, your left hand hanging down, and, at the same time, bring down your right knee upon the firelock, looking up, both hands, the left knee and heel in a line. Tell 1, 2, and quitting the firelock with the right hand, raise yourself up again, stepping back with your left foot, and keeping your body half faced to the right. Tell 1, 2, and, turning your right foot on the heel over the butt-end, bring your body to its proper front, letting both arms hang down by your sides.

IV. *Take up your firelocks!* 4 motions.

Turn your right foot upon the heel over the butt of the firelock, and set it down again, making a half face to the right; at the same time, extend your right arm a little to the right, the palm towards the front. Tell 1, 2, and step forwards with the left foot along the firelock, at the same time take hold of it with the right hand, about the middle of the barrel, with an outstretched arm, and a stiff body. Tell 1, 2, and raise yourself, and the firelock, bringing back the left foot; then tell 1, 2, and lift up your right foot, setting it at the inside of the butt, at the same time quitting the middle of the firelock with the right hand, and seizing it again briskly with the same hand, as high as the muzzle,

turning

turning the barrel towards your right shoulder, being then in the position directed in explanation the second.

V. *Rest your firelocks!*   3 motions.

Quitting the firelock with the right hand, seize it again briskly with the same hand, as low as you can without constraint. Tell 1, 2, and raise your firelock with the right hand, seizing it at the same time with the left hand, just above the feather-spring, the left arm across the stomach. Tell 1, 2, quit the firelock with the right hand, and take hold of it behind the lock, at the same time turning upon both heels, the right toe pointing to the right, the firelock and body being in the attitude, as in explanation the first.

VI. *Club your firelocks!*   3 motions.

Keeping your firelock fast in the left hand, cast it about close by your right shoulder, in the direction of the files, turning upon both heels, so as to bring your feet to their proper front, take hold of it at the same time with the right hand, as low as you can without constraint, the cock the height of the roller, the muzzle of the piece, and left thumb downwards, the lock outwards. Tell 1, 2, let go the left hand, and raising the firelock with the right, take hold of it again with the left hand at the small end of the stock, that hand being at the height of the waist-belt, bringing it near the left shoulder, the firelock being perpendicular. Tell 1, 2, and bring it upon your left shoulder, the left arm close to the body, the lock upwards; at the same time quitting it with your right hand, which is to hang down along your right side.

VII. *Rest your firelocks!*   3 motions.

With a brisk motion seize your firelock with the right hand even with the shoulder, raising the piece from your
shoulder

Chap. II.   *Military Discipline.*   23

shoulder at the same time, so as to bring it perpendicular. Tell 1, 2, let go your left hand, and sinking the firelock with the right, seize it with the left hand turned, near the lock, the thumb downwards, at the same time, turning the barrel outwards, the guard even with your roller. Tell 1, 2, and letting go the firelock with the right hand, turn it with the left, as in the club, bringing the butt-end downwards, and, turning on your heels, come to the rest.

VIII. *Secure your firelocks!*   3 motions.

Raising your firelock with your right hand, and quitting it with the left, come to the poise. Tell 1, 2, and, turning the barrel outwards, bring the firelock to the left side, the muzzle directly up; at the same time, seize the piece with the left hand at the swell, below the tail-pipe, sinking the right a little, so as to bring the right arm across the breast. Tell 1, 2, quit your right hand, and bring your firelock, with your left hand, under your left arm, the barrel downwards, the wrist the height of the waist-belt.

IX. *Shoulder your firelocks!*   3 motions.

With a quick motion bring the firelock from under your arm, raising the muzzle so as to bring it perpendicular, at the same time seizing it with the right hand under the lock, the left at the feather-spring, and at the height of the roller. Tell 1, 2, and quitting it with the left hand, bring the firelock with the right hand opposite to your left shoulder, the barrel outwards, at the same time placing the butt in the left hand, so as the thumb and fore-finger may be above the swell of it, the middle finger just upon the swell, and the other two fingers under the butt, the piece upright, but sunk so, as to bring the guard a little lower than the left breast, the elbows down, the butt close to the hip, the lock a little turned to the front. Tell 1, 2, and let the firelock fall upon the left shoulder, throwing back the right arm.

X. *Poise your firelocks!* 2 motions.

Seize the firelock with the right hand behind the cock, at the same time turning it with the left hand, so as to bring the lock upwards. Tell 1, 2, and raising the firelock from your shoulder, bring it before you, the lock turned outwards, letting your left hand fall down along the left side, your right arm, as far as the elbow, close to your body, by which means, the firelock is better supported.

XI. *Rest upon your arms!* 3 motions.

With your right hand sink your firelock close to your body, as low as you can without constraint, seizing it, at the same time, with your left, the height of your chin. Tell 1, 2, sink the firelock with the left hand, seizing it at the muzzle with the right. Tell 1, 2, and bring the butt to the ground, at the same time bringing up the left hand close under the right, the elbows down, and the firelock close to your body.

XII. *Draw your bayonets!* 2 motions.

Push forward the muzzle of your firelock with the left hand, at the same time seize the bayonet with the right hand. Tell 1, 2, and draw it out briskly, and, turning the point upwards, bring the socket just above the muzzle of the piece, the bayonet upright, the elbow down.

XIII. *Fix your bayonets!* 3 motions.

Push down the socket of your bayonet, as far as the notch will permit. Tell 1, 2, and turning it from you,
fix

fix it. Tell 1, 2, and bring down the right hand, upon the back of the left.

XIV. *Poife your bayonets!*   3 motions.

Raife your firelock with your right hand, as high as your forehead, at the fame time feizing it with the left hand as low down as poffible, without conftraint. Tell 1, 2, raife it with the left hand, and, at the fame time, quitting it with the right, feize it again with that hand, below the lock, the left hand at the feather-fpring. Tell 1, 2, and, quitting the left hand, let it fall along the left fide, remaining in the pofture defcribed in explanation the tenth.

XV. *Shoulder!*   2 motions.

Bring the firelock with the right hand oppofite to the hollow of the left fhoulder, turning the barrel outwards, and placing the butt in the left hand, as in explanation 9. Tell 1, 2, and let it fall on the left fhoulder, as in the faid explanation.

XVI. *Prefent your arms!*   3 motions.

As in explanation the firft.

XVII. *To the right!*   3 motions.

Bring the firelock to the recover, at the fame time, place the right heel four inches behind the left, the right toe pointing to the right. Tell 1, 2, and face upon both heels to the right. Tell 1, 2, and come down to the reft, placing the feet as directed in that attitude.

XVIII. *To the right!*   3 motions.

As the foregoing.

XIX.

XIX. *To the right-about!*   3 motions.

As the foregoing, only facing upon both heels to the right-about.

XX. *To the left!*   3 motions.

Bring the firelock to the recover, at the same time bring up the right heel to the ball of the left foot, the right toe pointing to the right. Tell 1, 2, and face upon both heels to the left. Tell 1, 2, and come down to the rest, as before.

XXI. *To the left!*   3 motions.

As the preceding.

XXII. *To the left about!*   3 motions.

As the preceding, only that you face upon both heels to the left-about.

XXIII. *Charge your bayonets!*   1 motion.

Step forward about eighteen inches with the left foot, bending the left knee, and at the same time seizing the butt with the right hand, placing the plate full in the palm of that hand, bring down the muzzle, so as the firelock may rest upon the left arm, almost level, and as high as your breast, the left elbow turned out towards the front, the fingers and thumb towards the lock.

XXIV. *Rest your bayonets on the left arm!*   3 motions.

Fall back with your left foot to its proper place, at the same time seizing the stock with your right hand, and bringing up the muzzle, come to the recover. Tell 1, 2, bringing the firelock directly before you, turn the lock

Chap. II. *Military Discipline.* 27

lock from you, the piece perpendicular. Tell 1, 2, and let go your left hand, sink the firelock, and at the same time seize the cock and steel with your left hand, the cock lying on your middle finger, and the lower joint of your thumb on the steel; keep both arms as low as possible without constraint, the butt between your thighs, and the bayonet pointing exactly to the left, as far from your shoulder, as the situation of both your arms, and the butt will permit.

XXV. *Rest your bayonets!* 3 motions.

Quitting the firelock with the left hand, seize the stock just below the tail-pipe, the thumb inwards. Tell 1, 2, bring the firelock to the recover. Tell 1, 2, and come down to the rest.

XXVI. *Shoulder!* 2 motions.

Turning upon both heels bring your feet to the proper front, and bring up your firelock briskly with both hands, over-against your left shoulder, at the same time placing the left hand under the butt, as in the second motion of explanation the 9th. Tell 1, 2, and come to the shoulder, as in the said explanation.

2d caution. *Take care to perform the platoon exercise!* or a ruffle on the drum.

XXVII. *Rear half files to the left double your front!*
2 motions.

The Officers in the rear, and the serjeants on the flanks, recover their espontons, and halberts, in two motions.

XXVIII.

XXVIII. *March!* 20 motions.

The three rear ranks, and the Officers, step forward at once with their left feet, and make eighteen ordinary paces of two feet each to the front, counting 1, 2, between each pace, and upon finishing the last, or eighteenth pace, with their right feet, immediately bring up the left, so as to be square, placing themselves on the left of their file-leaders: After the eighteenth pace, and the bringing up of the left feet, the Officers and serjeants order their arms in two motions.

XXIX. *Prime and load!* 21 motions.

1st. *Join your right hand to your firelock!* 1 motion.

Tell 1, 2, and,

2d, *Recover your firelock!* 1 motion.

As in explanation the first. Tell, 1, 2, and,

3d. *Open your pans!* 2 motions.

Stepping back with the right foot four inches behind the left heel, and facing full to the right, the left hand half way between the swell, and the feather-spring; bring back the butt of the firelock, the lock just below the right breast, the left arm pressed against the body, so as to support the piece, the muzzle of which is to be raised as high as the man's head in the rank before you; at the same time placing the ball of the right thumb behind the hammer, the fingers shut. Tell 1, 2, force it back, the elbows down. Tell 1, 2, and,

4th. *Handle your cartridge!* 1 motion.

Bringing down your right hand to your pouch, taking

ing out a cartridge with your two forefingers and thumb, the thumb towards the hip, and bring it to your mouth, the elbow a little turned up. Tell 1, 2, and,

5th. *Open your cartridge!* 1 motion.

By biting off the top of the cartridge paper, so as to come at the powder, and placing instantly your thumb upon the mouth of the cartridge, bring it opposite to, and just above the pan, the cartridge perpendicular. Tell 1, 2, and,

6th. *Prime!* 2 motions.

By placing the thumb on the edge of the pan, and, turning up your hand, shake carefully some of the powder into it, covering again immediately the mouth of the cartridge with the thumb. Tell 1, 2, and place the two last fingers behind the hammer, the cartridge being upright.

7th. *Shut the pan!* 2 motions.

By a short, and quick motion, with your two last fingers. Tell 1, 2, and pushing down the butt, cast back the muzzle of the firelock, catching it in the hollow of the right hand, letting the firelock slip through the left hand, till it comes to the swell, near the tailpipe; pressing the left hand against the waist-band, the butt opposite to the left toe: the piece, in turning, must be kept close to the body, the cartridge still covered with the thumb, and close to the muzzle, in a line with the barrel, the right elbow turned down. Tell 1, 2, and,

8th. *Load with cartridge!* 2 motions.

By putting it into the barrel, and shaking out the powder, push the ball into the muzzle with the forefinger; then tell 1, 2, and seize immediately the butt end

end of the rammer, with your thumb, and forefinger, the thumb upwards, the other fingers clenched. Tell 1, 2, and,.

9th. *Draw your rammer!* 3 motions.

As far as you can, catching it inftantly with your right hand, the thumb turned downwards. Tell 1, 2, and clear it of the pipes, turning it immediately, and placing the butt-end of it againft your fword-belt, fhorten it within three inches of the butt end, by flipping down your hand, the rammer in the fame direction as the barrel. Tell 1, 2, and bring the butt end of it into the muzzle, upon the cartridge, the elbows clofe. Tell 1, 2, and,

10th. *Ram down your cartridge!* 3 motions.

Seize the rammer in the middle, drive it down with good force, catching it quickly afterwards at the muzzle, the thumb turned downwards, and recovering it half-way. Tell 1, 2, and draw it entirely out of the barrel, turning it, and placing the fmall end againft your fword-belt, fhorten it by flipping down your hand within twelve inches of the end. Tell 1, 2, and bring the fmall end into the firft pipe, conducting it down with the finger and thumb, through the fecond pipe, the finger and thumb pointing upwards; then tell 1, 2, and,

11th. *Return your Rammer!* 1 motion.

By placing the forefinger on the butt of the rammer, with a quick and ftrong motion force it quite down, at the fame time raife the firelock with the left hand, bringing immediately the right hand under the lock, the left hand at this time flipping down to the feather-fpring, the cock at the height of the waift belt. Tell 1, 2, and,

12th. *Shoulder !* 2 motions.

By facing to the left, and bringing up your right heel within four inches of the left, at the same time bring the firelock close and short about, opposite to the left shoulder, the barrel outwards, and placing the butt in the left hand, close to the hip. Tell 1, 2, and drawing the elbows briskly back, bring the piece to the left shoulder, throwing back the right hand, as in explanation the ninth.

N. B. *The firing quick depends upon the quick loading, and that chiefly upon the dexterity of drawing, and returning the rammer; that part therefore of the Exercise, requires great practice and attention.*

XXX. *As front rank, make ready !* 3 motions.

Join your right hand, and recover your firelock in two motions, as in explanation the first, placing the thumb upon the cock, the forefinger behind the guard. Tell 1, 2, and stepping back with the right foot in a direct line, kneel upon the right knee, the right toe turned inwards, and heel upright, the perpendicular line of the body falling about twelve inches behind the left heel, the body by that means properly poised, and upright; the butt end of the firelock placed at the same time upon the ground in a line with the left heel: upon coming down to the kneel, you cock the firelock.

XXXI. *Present !* 1 motion.

Bring down the muzzle of your piece with both hands, throwing forward your left hand, as far as the swell of the stock under the barrel, and placing the butt end in the hollow betwixt your right breast and shoulder, pressing it close to you, at the same time taking your right thumb from the cock, and placing the forefinger upon the tricker, both arms close to your body, taking good aim, by leaning the head to the right, and looking along the barrel.

XXXII. *Fire!* 19 motions.

1st. Draw the tricker strongly, and at once, with your fore-finger, and immediately on having fired, rise from the kneel, bringing the right heel four inches behind the left foot, at the same time bring back the firelock, as in the third motion of the twenty-ninth explanation, the right thumb upon the cock.

2d. *Half-cock your firelock!* 1 motion.

By straining the tumbler to the half-bent with your right thumb, bringing down your right elbow at the same time, by that motion adding force to it. Tell 1, 2, and,

3d. *Handle your cartridge!* 1 motion.
4th. *Open your cartridge!* 1 motion.
5th. *Prime!* 2 motions.
6th. *Shut your pan!* 2 motions.
7th. *Load with cartridge!* 2 motions.
8th. *Draw your rammer!* 3 motions.
9th. *Ram down your cartridge!* 3 motions.
10th. *Return your rammer!* 1 motion.
11th. *Shoulder!* 2 motions.

As in explanation 29, making in all, 19 motions, and counting 1, 2, between each.

XXXIII. *As center-rank, make ready!* 3 motions.

As the foregoing, only instead of stepping back three feet, and kneeling, they step back with the right feet
eighteen

Chap. II.   *Military Difcipline.*   33

eighteen inches, in a direct line to the rear, by that means bringing their feet juft behind the right feet of the front rank, cocking the firelock, and keeping it upright at the recover, right elbow down.

XXXIV.   *Prefent!*   1 motion.

Bringing down the firelock, as in explanation the thirty-firft, only a little to the right of the front rank.

XXXV.   *Fire!*   19 motions.

As in explanation the thirty-fecond.

XXXVI.   *As rear rank, make ready!*   3 motions.

As in explanation the thirty-third, only this rank fteps to the right with the right foot, inftead of falling back with it, till their toes touch the hinder part of the left heels of their right hand men, at the fame time bending their right knees a little, fo that their bodies may be oppofite to the intervals of their file-leaders, and of the files upon their right, the firelock held in the fame attitude as in explanation the thirty-third.

XXXVII.   *Prefent!*   1 motion.

As in explanation the thirty-fourth, only bring down the firelock between that of the file-leader, and that of the right hand file.

XXXVIII.   *Fire!*   19 motions.

As in explanation the thirty-fecond.

XXXIX.   *Rear ranks clofe to the front!*   3 motions.

The Officers advance their arms in three motions, as do the ferjeants upon the flanks.

D   XL.

XL. *March!* 10 motions.

The center and rear ranks, as also the rank of Officers, step off together, with the left feet, the center rank make five paces, and bring up their right feet; the rear rank, and the rank of Officers, make ten paces, and bring up their left feet: the ranks are then at one pace, or two feet asunder, except the Officers, who are at the same distance from the rear, as before they moved.

XLI. *Make ready!* 3 motions.

The three ranks make ready together, the front rank observing the directions in explanation the thirtieth; the center rank, the directions in explanation the thirty-third; and the rear rank, those in explanation the thirty-sixth.

XLII. *Present!* 1 motion.

As in explanation the thirty-first, thirty-fourth, and thirty-seventh.

XLIII. *Fire!* 1 motion.

Having fired, rise from the ground, the three ranks coming to the recover, as in explanation the first.

XLIV. *Charge your bayonets!* 1 motion.

The front rank only charges their bayonets, as in explanation the twenty-third. The serjeants on the flanks of that rank, charge their halberts in like manner; the other two ranks remain with recovered arms.

XLV. *Recover your arms!* 1 motion.

The front rank falling back with the left feet, recover their arms, as in explanation the first.

XLVI. *Rear ranks, to your proper distance!* 3 motions.

The center and rear rank, as also the rank of Officers, go to the right-about, as in explanation the nineteenth.

XLVII. *March!* 10 motions.

The rank of Officers, and the rear rank, step off with their right feet, and having marched ten paces, halt on the left feet; when the rear rank makes its sixth pace, the center rank steps off with the left feet, and halts at the fifth pace, when the rear rank does.

XLVIII. *Halt!* 3 motions.

The ranks which marched, bring their right feet opposite to the left toe, and go to the left-about, as in explanation the twenty-second; the Officers and serjeants order their arms in three motions, at the same time that they face.

XLIX. *Shut your pans!* 6 motions.

Half-cock the firelock, by straining the tumbler to the half-bent, with your right thumb, bringing down your right elbow. Tell 1, 2, and bring down your firelock, as when you handled your cartridge. Tell 1, 2, and take the wiping-cloth out of your pouch, bringing it up to the pan. Tell 1, 2, clean the pan, and bring the fingers behind the hammer. Tell 1, 2, and shut the pan, as when you held the cartridge.

Tell 1, 2, and cast back the muzzle, as if to charge.

L. *Clean your bayonets!* 2 motions.

Clean your bayonet with your wiping-cloth, and return the cloth into your pouch. Tell 1, 2, and bring the firelock upright, raising it with your left hand, at the same time bringing the right hand under the lock, as in explanation the twenty-ninth.

LI. *Shoulder!* 2 motions.

As in exlanation the twenty-ninth.

LII. *Rear half files, as you were!* 2 motions.

The Officers in the rear, and the rear half files, face to the right on their left heels in two motions, the Officers, and the serjeants of the rear half files, recovering their arms at the same time.

LIII. *March!* 18 motions.

They proceed to their former distance, making the last pace with the left feet.

LIV. *Halt!* 3 motions.

They all come to the left about in three motions, as in explanation the forty-eighth, the Officers and serjeants ordering their arms at the first and last motions.

LV. *Front half files to the right double your rear!* 3 motions.

They go to the right-about in three motions, the serjeants upon the flanks recovering their halberts at the

first

first and last motions: the Pioneers, who had stood fast till now, go to the right about, and come about again with the front half files.

LVI. *March!* 18 motions.

They march eighteen paces, inclining to their left, till they are opposite to the intervals, and double the rear half files upon the right, facing full to the left of the battalion.

LVII. *Halt!* 2 motions.

Place the right heel four inches behind the left heel. Tell 1, 2, and face upon both heels to the front; the serjeants order their halberts at the same time.

LVIII. *Front half files, as you were!* 2 motions.

The serjeants recover their arms.

LIX. *March!* 20 motions.

They proceed to their former ground in eighteen paces; the serjeants having counted 1, 2, order their arms in two motions.

N. B. The pioneers march at the same time that the front half files do.

Third caution to the Officers, or a roll upon the drum.

Officers, recover your arms! or *a flam.* 3 motions.

All the Officers and serjeants recover their arms in two motions; the serjeants in the front, face to the right and left outwards; at the same time, those on the flanks, face to the right-about in 3 motions; the drummers and musick face likewise to the right and left outwards, except the orderly drummers.

March ! or the *troop.*   43 motions.

The Officers stepping off together with their left feet, and observing to keep a straight rank, as they march, move up through the files in forty-three paces to their ground in the front; the serjeants march to their posts in the rear, and on the flanks; the drummers having marched so as to be clear of the flanks, face together, and proceed till they are in a line with the front rank.

Halt ! or a *flam.*   3 motions.

Those who were faced, come to their proper front.

Order your arms ! or a *flam.*   2 motions.

The Officers order their arms in 2 motions, as do the serjeants.

4th caution.   *Rear half files, take care to double your front !* or a *ruffle.*

LX. *Rear half files, to the left double your front !*
2 motions.

The serjeants of those half files, as well as the serjeants in the rear, recover their arms in 2 motions.

LXI. *March !*   20 motions.

As in explanation the twenty-eighth, the serjeants moving up at the same time: as soon as these last are
come

come to the ground, they are to order their halberts in two motions.

LXII. *Left hand grand divifions, by the fide ftep double your files to the right!* 4 motions.

The ferjeants and men of the grand divifions which are to double, make three paces backwards, beginning with their left feet; and after the third ftep, they bring back their right feet in a line with the left, their right toe pointing to the right; the Officers make only two paces, and then bring back the left feet: the Officers and ferjeants recover their arms at the two firſt paces.

LXIII. *March!*

Officers, ferjeants, and men, ftep off together, with the left feet, and bring the left heel oppofite to the right toe, the left feet pointing to the front; the fecond ftep is made by moving the right feet in a direct line to the right, and fo on alternately, till the divifions have doubled: the Officers, ferjeants, and men, muſt take care to raife the fame feet together, that they may not open, or clofe their files, and without facing to the right: the laſt ftep muſt be made with the right feet.

LXIV. *Halt!* 2 motions.

They all turn their right feet to the proper front, the Officers and ferjeants ordering their arms in 2 motions.

N. B. They will take care to cover their files immediately.

LXV. *Grand divisions, to the left as you were!* 2 motions.

The Officers, serjeants, and men, who had doubled, turning on their left heel, point the left toe to the left, the Officers and serjeants recovering their arms.

LXVI. *March!*

They proceed by stepping off together, with the right feet, bringing the right heel opposite to the left toe, the right feet pointing to the front, and the left feet pointing directly to the left; the next step is made by stepping with the left feet in a direct line to the left, and so on, till they have marched clear of the divisions they had doubled.

LXVII. *Halt!* 4 motions.

Bring the left toe to its proper front. Tell 1, 2, and, stepping off with the left feet, the ranks make three paces to the front; the Officers make only two paces, and telling 1, 2, they and the serjeants order their arms together, in 2 motions: the whole will take care to dress instantly with the divisions which stood fast.

LXVIII. *Right hand grand divisions, by the side step double your files to the left!* 4 motions.

The same as in explanation the sixty-second, only beginning with the right foot, and pointing the left toe to the left.

N. B. The pioneers do not move.

LXIX. *March!*

Stepping off together, with their right feet, and bringing that heel oppofite the left toe; then ftepping with the left feet in a direct line to the left, they proceed till the whole divifions have doubled, the laft pace made with the left feet.

LXX. *Halt!* 2 motions.

They turn the left feet to the proper front, the Officers and ferjeants ordering their arms, and immediately dreffing their ranks.

LXXI. *Grand divifions, to the right as you were!* 2 motions.

The Officers, ferjeants and men, who had doubled, turning on the right heel, point their right toe to the right, the Officers and ferjeants recovering their arms.

LXXII. *March!*

They proceed till they have marched clear of the divifions which ftood faft.

LXXIII. *Halt!* 4 motions.

They bring the right toe to the proper front. Tell 1, 2, and move to the front, the Officers making two, and the ferjeants and men, three paces; then, telling 1, 2, the Officers and ferjeants order their arms in 2 motions, dreffing the ranks immediately.

LXXIV

LXXIV. *Rear ranks, close to the front!* 3 motions.

The Officers and the serjeants advance their arms in 3 motions. The drummers fall in between the rear rank, and the serjeants.

LXXV. *March!* 10 motions.

The center rank makes five paces; the rear rank, and rank of serjeants, make ten paces.

LXXVI. *Upon the center, wheel to the right about!*
3 motions.

The two grand divisions upon the right of the battalion, with the Officers of those divisions, and right hand division of the grenadier company, face to the right about in 3 motions.

LXXVII. *March!*

The battalion wheels to the right-about upon its center, as does the grenadier company; but the last, very slowly.

N. B. The pioneers stand fast.

LXXVIII *Halt!* 4 motions.

The two grand divisions of the battalion, and the division of grenadiers, which had faced, stand square: then tell 1, 2, and come to the right-about.

LXXIX. *Upon the center, wheel to the left-about!*
3 motions.

The two grand divisions upon the left of the battalion,

talion, and the left hand division of grenadiers, face to the right-about in 3 motions.

LXXX. *March!*

The battalion wheels to the left-about upon its center, as does the grenadier company; only this last, wheels very slowly.

LXXXI. *Halt!* 4 motions.

The divisions which had faced, stand square: then tell 1, 2, and come to the right-about.

LXXXII. *Rear ranks, take your proper distance!* 3 motions.

The serjeants, and the rear ranks, face to the right-about, as in explanation the forty-sixth.

LXXXIII. *March!* 10 motions.

The serjeants in the rear, and the rear rank, march ten paces; and the center rank, five paces to the rear, as in explanation the forty-seventh.

LXXXIV. *Halt!* 3 motions.

They come to the left-about in 3 motions, the Officers and serjeants ordering their arms at the same time.

LXXXV. *Poise your firelocks!* 2 motions

As in explanation the tenth.

LXXXVI.

LXXXVI. *Rest upon your arms!*   3 motions.

As in explanation the eleventh.

The commanding Officer advances to the reviewing General, and takes his orders with regard to what should be done next. If the General should order the regiment to go through the firings, as formerly directed by His Royal Highness the Duke, after the battalion is poised, and shouldered, the Officers are to take post according to His Royal Highness's orders of the 18th of *April*, 1756.

LXXXVII. *Unfix your bayonets!*   3 motions.

Strike up the bayonet strongly with your right hand, elbow down. Tell 1, 2, and turn it to the right. Tell 1, 2, and bring it off the piece; and pushing forward the muzzle at the same time, enter the point into the scabbard, throwing up the head, after pointing, to its proper position.

LXXXVIII. *Return your bayonets!*   2 motions.

Thrust it home. Tell 1, 2, and bring your right hand over your left. Elbows close.

LXXXIX. *Poise your firelocks!*   3 motions.

As in explanation the fourteenth.

XC. *Shoulder.*   2 motions.

As in explanation the fifteenth.

XCI. *Rear half files, as you were!*   2 motions.

As in explanation the fifty-second.

XCII.

XCII. *March!* 18 motions.

As in explanation the fifty-third.

XCIII. *Halt!* 3 motions.

As in explanation the fifty-fourth.

CHAP. III.

The grenadier exercise, with an explanation, beginning when they are under arms, viz. the firelock shouldered.

HE grenadiers must observe the same directions, for the position under arms, as are given to those of the battalion: but they having an occasion for match for their granades, the match must be placed in the left hand, one end of it betwixt the first and second fingers, and the other between the two last, both ends standing a finger's length above the back of the hand. The rest of the match is to hang down by the inside of the butt-end of the firelock. The match is not to be lighted without express orders for it.

When the grenadiers stand in a body with the men of the battalion, they must then perform the same motions that they do, because they do not then meddle with their granades; and, consequently, there must be no difference either in the time or motions of the exercise of the grenadiers, and those of the battalion, except in the use of the slings and granades. I shall therefore refer to the explanation of the battalion exercise, except in those things which peculiarly belong to the grenadiers.

WORDS

WORDS of COMMAND.

Grenadiers, take care!

Though this is not reckoned a word of command, but only looked upon as a warning, to prepare them for their exercife; yet, whenever the grenadiers exercife apart from the battalion, they have annexed two motions to it, which, it muft be owned, have a very good effect, both on the fpectators and performers, by preparing the latter to go through their exercife with life, vigour, and exactnefs, in which the principal beauty of exercife confifts. The motions are as follows: firft, the grenadiers bring up their right hands brifkly, to the front of their caps. Then tell 1, 2, and bring them down with a flap on their pouches, with all the life imaginable; in which motions, neither their heads, bodies, nor firelocks, are to move.

I. *Make ready!* 3 motions.

This is performed, as in explanation the forty-firft, of the foot exercife.

II. *Prefent!* 1 motion.

As in explanation the thirty-firft.

III. *Fire!*

Having fired, come up to your recover, as directed in the fecond motion of explanation the firft.

IV. *Handle your flings!* 1 motion.

Quit the firelock with the left hand, extend your fling to the left, the thumb upwards, keeping the
fling

sling in a line with the firelock; and remain so till the next word of command.

V. *Sling your firelocks!* 3 motions.

Bring the sling with the left hand opposite to the right shoulder, and the firelock with the right hand opposite to the left shoulder, by crossing of both hands at the same time, bringing the left hand within the right, keeping the muzzle directly up, the barrel to the left, and the right hand just under the left elbow. Tell 1, 2, bend the firelock back, and bring the sling over your head, placing it just above your right shoulder, and the firelock opposite to the point of the left. Then tell 1, 2, draw the sling with your left hand, and let go the firelock with the right at the same time, that it may hang by the sling on the right shoulder, the muzzle upwards, and dropping both hands down by your sides at the same instant of time.

VI. *Handle your matches!* 3 motions.

Bring both hands directly before you with half-stretched out arms, about the height of your shoulders, taking hold of the lower end of the match at the same time with the right hand, placing the thumb under, and the two fore-fingers above. Tell 1, 2, and bring the match with the right hand over the back of the left, placing it between the thumb and two fore-fingers of the said hand. Then tell 1, 2, thrust out your left hand with the match straight forward, by extending the arm at full length, and, at the same time, bring your right hand down to your right side.

VII. *Handle your Granades!* 3 motions.

Keep your left hand extended to the front, as before, and face nimbly to the right on the left heel, stretching out your right arm, at the same time, the height of your shoulder, pointing directly to the rear. Tell 1, 2, and clap your right hand briskly on your pouch, sezing (if there should be occasion) your granade. Then tell 1, 2, and bring up your right hand to its former position, placing the thumb against the fuze, and continue in this position till the following word of command.

VIII. *Open your fuze!* 3 motions.

Keep your left hand extended to the front, and bring the granade with your right hand to your mouth. Tell 1, 2, and open the fuze with your teeth. Then tell 1, 2, thrust your arm, nimbly, from you to its former place.

IX. *Guard your fuze!* 1 motion.

Cover the fuze with your thumb, without making any other motion.

X. *Blow your matches!* 2 motions.

Bring the match with your left hand before your mouth; then tell 1, 2, and blow it off with a strong blast, thrusting back your hand, at the same time, to its former place.

XI. *Fire and throw your granades!* 3 motions.

Meet the granade with your left hand, opposite to your right thigh, inclining your body to the right side, bending the right knee, and keeping the left stiff, and fire the fuze at the same time. Tell 1, 2, slowly, that the fuze may be well lighted, and throw the granade with a stiff arm, stepping forward, at the same time, with the right foot, placing it in a line with the left, extending both arms in a direct line to the front, keeping the left uppermost, and the body upright. Then tell 1, 2, and bring your right hand down to your side, keeping your left in its former position.

XII. *Return your Matches!* 3 motions.

Bring both hands before you, as directed by the first motion of explanation the tenth. Tell, 1, 2, and bring the match back to its former place, between the two last fingers of the left hand. Then tell 1, 2, and let both hands fall down by your sides.

XIII. *Handle your slings!* 3 motions.

Seize the sling with both hands at the same time, taking hold of it with the right hand about the middle, and as low as you can reach, without bending your body, with the left. Tell 1, 2, and with the left hand bring the butt forward, slipping your left elbow under the firelock, by bringing of it between the firelock and the sling; take hold of the firelock, at the same time, with the left hand, letting the stock lie between the thumb and the fore-finger, the butt end pointing a little to the left, with the barrel upwards. Then tell 1, 2, bring the firelock to lie on the left shoulder, and the sling on the right, the barrel upwards, and the butt-end pointing directly to the front, keeping the firelock to a true level.

XIV. *Recover your firelocks!* 1 motion.

Bring the firelock before you, seizing it briskly with the right hand under the lock, and with the left

at

at the feather-spring, turning the barrel, at the same time, inwards, as in the second motion of explanation the first, in the manual exercise, the right thumb upon the cock.

XV. *Half-cock your firelocks!* 1 motion.

Half-cock the firelock, by straining the tumbler to the half bent, with your right thumb, bringing down your right elbow, at the same time, by that motion adding force to it.

XVI. *Shoulder your Firelocks!* 2 motions.

Quit the firelock with the left hand, and bring it with the right opposite to your left shoulder, placing the butt, at the same time, in the left hand, as is directed in the second motion of explanation the ninth.

XVII. *Shut your Pans!* 3 motions.

Bring up your right hand, and place the two forefingers of it behind the hammer. Tell 1, 2, and shut the pan. Tell 1, 2, and bring your right hand down to its former position by the right side.

Note, As the rest of the exercise, is the same with the Battalion, I shall only set set down the words of command, with the explanitions referred to.

1. *Rest your firelock!* Explanation 1.
2. *Order your firelocks!* Explan. 2.
3. *Ground your firelocks!* Explan. 3.
4. *Take up your firelocks!* Explan. 4.
5. *Rest your firelocks!* Explan. 5.
6. *Club your firelocks!* Explan. 6.
7. *Rest your firelocks!* Explan. 7.
8. *Secure your firelocks!* Explan. 8.
9. *Shoulder your firelocks!* Explan. 9.

10. *Poise your firelocks!* Explan. 10.
11. *Rest upon your arms!* Explan, 11.
12. *Draw your bayonets!* Explan. 12.
13. *Fix your bayonets!* Explan. 13.
14. *Poise your bayonets!* Explan. 14.
15. *Shoulder!* Explan. 15.
16. *Present your arms!* Explan. 16.
17. *To the right!* } Explan. 17.
18. *To the right!* }
19. *To the right-about!* Explan. 19.
20. *To the left!* } Explan. 20.
21. *To the left!* }
22. *To the left-about!*
23. *Charge your bayonets!* Explan. 23.
24. *Rest your bayonets on the left arm!* Explan. 24.
25. *Rest your bayonets!* Explan. 25.
26. *Shoulder!* Explan. 26.
27. *Rear-half-files, to the left* } Explan. 27.
    double your front! }
28. *March!* Explan. 28.
29. *Prime and load!* Explan. 29.
30. *As front rank, make ready!* Explan. 30.
31. *Present!* Explan. 31.
32. *Fire!* Explan. 32.
33. *As center rank, make ready!* Explan. 33.
34. *Present!* Explan. 34.
35. *Fire!* Explan. 35.
36. *As rear rank, make ready!* Explan. 36.
37. *Present!* Explan. 37.
38. *Fire!* Explan. 32.
39. *Rear ranks, close to the front!* Explan. 39.
40. *March!* Explan. 40.
41. *Make ready!* Explan. 30, 33, and 36.
42. *Present!* Explan. 31, 34, and 37.
43. *Fire!* Explan. 43.
44. *Charge your bayonets!* Explan. 44.
45. *Recover your arms!* Explan. 45.
46. *Rear ranks, to your proper distance!* Explan. 19.
47. *March!*

Chap. III. *Military Discipline.*

47. *March!* Explan. 47.
48. *Halt!* Explan. 48.
49. *Shut your pans!* Explan. 49.
50. *Clean your bayonets!* Explan. 50.
51. *Shoulder!* Explan. 29.
52. *Rear half files, as you were!* Explan. 52.
53. *March!* Explan. 53.
54. *Halt!* Explan. 48.
55. *Front half files, to the right double your rear!* } Explan. 55.
56. *March!* Explan. 56.
57. *Halt!* Explan. 57.
58. *Front half files, as you were!* Explan. 58.
59. *March!* Explan. 39.
60. *Rear half files, to the left double your front!* } Explan. 60.
61. *March!* Explan. 28.
62. *Left hand grand divisions, by the side step double your files to the right!* } Explan. 62.
63. *March!* Explan. 63.
64. *Halt!* Explan. 64.
65. *Grand divisions, to the left as you were!* } Explan. 65.
66. *March!* Explan. 66.
67. *Halt!* Explan. 67.
68. *Right hand grand divisions, by the side step double your files to the left!* } Explan. 68.
69. *March!* Explan. 69.
70. *Halt!* Explan. 70.
71. *Grand divisions, to the right as you were!* } Explan. 71.
72. *March!* Explan. 72.
73. *Halt!* Explan. 73.
74. *Rear ranks, close to the front!* Explan. 74.
75. *March!* Explan. 40.

76. *Upon the center, wheel to the right-about!* } Explan. 76.
77. *March!* Explan. 77.
78. *Halt!* Explan. 78.
79. *Upon the center, wheel to the left-about!* } Explan. 79.
80. *March!* Explan. 80.
81. *Halt!* Explan. 81.
82. *Rear ranks, take your proper distance!* Explan. 46.
83. *March!* Explan. 47.
84. *Halt!* Explan, 48.
85. *Poise your firelocks!* Explan. 10.
86. *Rest upon your arms!* Explan. 11.
87. *Unfix your bayonets!* Explan. 87.
88. *Return your bayonets!* Explan. 88.
89. *Poise your firelocks!* Explan. 14.
90. *Shoulder!* Explan. 15.
91. *Rear half files, as you were!* Explan. 52.
92. *March!* Explan. 53.
93. *Halt!* Explan. 54.

The grenadiers having distinct words of command for the firing and throwing of their granades, besides those above-mentioned, I shall therefore set them down in the order, as they are to follow.

Take heed!

I. *Make ready!*

As in explanation the forty-first, of the battalion exercise.

II. *Present!*

As in explanation the thirty-first.

III. *Fire!*

Having fired, come up to your recover, as above directed, and remain in that position, till the following command is given.
Take

Take care to fire and throw your granades, at 3 words of command!

I. *Make ready!*

This muſt be done, by performing all the motions in the grenadier exerciſe, from explanation the fourth to the ninth incluſive.

II. *Blow your matches!*

As in explanation the tenth, of the ſaid exerciſe.

III. *Fire and throw your granades.*

This is done, as in explanation the ſecond. After which they are to go on, till they have performed the 16th word of command.

The end of the grenadier exerciſe.

CHAP. IV.

General rules for wheeling.

THE circle is divided into four equal parts.

Wheeling to the right or left, is only a quarter of the circle.

Wheeling to the right or left-about, is one half of the circle.

When you wheel to the right, you are to close to the right so near as to touch your right hand man, but without pressing him, and to look to the left in order to bring the rank about even.

When you wheel to the left, you are to close to the left and look to the right, as above directed.

This rule will serve for all wheeling by ranks; as when a battalion is marching by sub-divisions with their ranks open, then each rank wheels distinctly by itself when it comes to the ground on which the rank before it wheeled, but not before.

It will likewise serve as a rule for the front rank in all wheelings, whether that of the whole battalion, or grand, or sub-divisions; but the rear ranks, when they are closed forward, being to wheel directly in the rear of, and at the same time with, the front rank, must incline a little to the left when they wheel to the right, in order to keep directly in a line with their file-leaders. The same rule must be observed by the rear ranks when they wheel to the left, by inclining a little to the right, for the reason above-mentioned, that of keeping in a line with their file-leaders.

In wheeling, the men are to take particular care neither to open nor close their ranks, and to carry their firelocks high and firm on their shoulders.

Chap. IV. *Military Discipline.*

In wheeling, the motion of each man is quicker or slower according to the distance he is from the right or left, thus; when you wheel to the right, each man moves quicker than his right hand man; and, in wheeling to the left, each moves quicker than his left hand man; the circle that every man wheels being larger, according to the distance he is from the hand he wheels to, as may be seen by describing of several circles within one another at two feet distance from each, which is about the space every man is supposed to take up.

I. *To the right wheel by division! March!*

At this command, they all step forward with their left feet, and wheel to the right a quarter of the circle.

The right hand man of the front rank of every division must turn on his right heel, without taking it out of its place; and, casting his eye to the left, bring his body and left foot about with the rank, according as it moves quicker or slower. All the men in the front rank are to cast their eyes to the left in the same manner, that they may neither advance before nor keep behind their left hand men; but to govern their steps in such a manner, that, by adding to, or abating from them, they may keep their bodies in a direct line with the left hand man of their division; by the due observance of which rule, the front rank will be always kept straight in wheeling, and, consequently, contribute a great deal towards the rear ranks doing the same; but unless the front rank wheels straight, it is impossible that the rear ranks should.

Let them observe further, as is directed by the general rules for wheeling.

All wheelings are to be done slow; and even those men on the extreme part of the ranks, are not to exceed a moderate pace.

The

The Officers are to wheel on the head of their several divisions, and the Serjeants on the flanks, and in the rear.

As soon as the divisions have wheeled a quarter of the circle, the Major is to proceed to the next word of command.

II. *Halt !*

At this they are to stand, and immediately dress their ranks and files.

III. *To the right, wheel ! March !*

All the divisions wheel a quarter of the circle again to the right, which brings the battalion to face to the rear. The directions, in explanation one, must be punctually observed.

IV. *Halt !* As in explanation the second.

V. *To the right, wheel ! March !*

As in explanation the first and third, by which they face to the left of the battalion.

VI. *Halt !* As in explanation the second and fourth.

VII. *To the right ! wheel ! March !*

This wheel completes the circle at four times, and brings them to their proper front.

VIII. *Halt !* As in explanation the second is directed.

IX. *Wheel to the right about!   March!*

By this word of command, all the divisions are to wheel one half of the circle to the right.

X. *Halt!*   As in explanation the second.

XI. *Wheel to the right about!   March!*

This wheel completes the circle at twice, and brings them to their proper front.

XII. *Halt!*   As in explanation the second.

XIII. *To the left, wheel!   March!*

The divisions are to wheel to the left a quarter of the circle, as they did before to the right; with this difference, that the left hand man of the front rank of each division must keep his left heel in its place, as the right hand men did their right heels when they wheeled to the right, and, by casting his eye to the right, bring his body and right foot about with the rank, according as it shall move quick or slow.

All the men of the front rank are to cast their eyes to the right, as they did before to the left, and for the same reason to keep their bodies in a straight line with the right hand man, that the rank may be kept even in wheeling.

The rear ranks are to incline a little to the right when they wheel to the left; as they did to the left when they wheeled to the right, observing further, as in explanation the first is directed, as also the general rules for wheeling.

XIV. *Halt!*   As in explanation the second.

XV.

XV. *To the left! wheel! March!* As in explanation the thirteenth is directed.

XVI. *Halt!* As in explanation the second.

XVII. *To the left, wheel! March!* As in explanation the thirteenth.

XVIII. *Halt!* As in explanation the second.

XIX. *To the left, wheel! March!*

As in explanation the thirteenth, which completes the circle at four times.

XX. *Halt!* As in explanation the second.

XXI. *Wheel to the left about! March!*

This wheel is one half of the circle to the left.

XXII. *Halt!* As in explanation the second.

XXIII. *Wheel to the left about! March!*

This wheel brings the divisions to their proper front.
When you would shorten them, it may be done by wheeling only twice to the right, and then to the right about, and the same to the left.

XXIV. *Halt!* As in explanation the second.

XXV. *Rear ranks, take your proper distance!*

The Serjeants, and the rear ranks, face to the right about, as in explanation the forthy-sixth.

XXVI.

XXVI. *March!*

The Serjeants in the rear, and the rear rank, march ten paces; and the center rank, five paces, to the rear, as in explanation the forty-seventh.

XXVII. *Halt!*

They come to the left about in 3 motions, the Officers and Serjeants ordering their arms at the same time.

CHAP. V.

Directions for passing in review.

ARTICLE I.

BEFORE the Regiment is to pass in review, the Companies should be drawn out, and a strict inspection made into the mens arms, ammunition, cloaths and accoutrements, and a report made of the same, by the Officer commanding each Company, to the Colonel or Officer commanding the Regiment, that he may know the true state of the whole, and give necessary directions in time, for the repairing of such things as may be then out of order.

If time and place will admit of it, it is usual for the Colonel to make this inspection himself along with the other Officers; or at least to order either his Lieutenant-Colonel or Major to do it, and to make him a report of the whole.

ARTICLE II.

The Regiment must be told off into front and rear half files; into sixteen platoons, eight sub-divisions, and four grand divisions: the Officers also are to be appointed to their proper posts; the Colours to be sent for, and the Regiment formed six deep, before the General comes; that, after he has viewed them standing.

Chap. V. *Military Discipline.* 63

ing, they may immediately proceed to the exercise, or whatever he shall be pleased to order,

Before the General Officer appears, the Major is to order the Regiment, 1st, to *Poise their firelocks*; 2dly, *Rest upon their arms*; 3dly, *Draw their bayonets*; 4thly, *Fix their bayonets*; 5thly, *Poise their firelocks*; and, 6thly, *Shoulder*.

N. B. It it be only a Major-General, who reviews, he is to be received without fixing of bayonets.

The General Officer being within twenty yards of the right flank of the battalion, the Major orders,

Present your arms!

and then takes his post on the right of the battalion, as the Adjutant does on the left, dressing with the front rank, and saluting the General, as he passes, with their swords.

As the General passes along the front, the Officers are to salute him with their espontons, if his rank be such, as to entitle him to that compliment; and must time it in such a manner, that each may just finish his salute, and pull off his hat when he comes opposite to him. The Ensigns who carry the colours are to drop them, if the General is to be saluted with colours, bringing the spear pretty near the ground, just when the Colonel drops the point of his half-pike, pulling off their hats at the same time, and not to raise the colours till he has passed them.

The drummers, at the same time, are to beat their two ruffles, three ruffles, or march, according to the General's rank.

As soon as the Major has saluted, he will post himself at about a hundred yards before the center of the Regiment, that he may be ready to order the men to face when the General goes round the battalion: and as it is impossible for the words of command to be distinctly heard by the whole, when the drums are

beating,

beating, the drummers should have directions to cease as soon as the General comes to the left flank of the battalion (supposing he began at the right) and not to begin beating till the word of command is given to face to the left; and when he comes to the left flank of the rear rank, they should cease again till the battalion has faced a second time to the left. The same rule should be observed when he comes to the right flanks of the rear and front ranks.

If the above directions are duly observed, the facings, which, on these occasions, are generally very ill performed, may be done with the utmost exactness. It is therefore incumbent on the Major, if he would shew the Regiment to advantage, not to neglect this precaution; since the performing of the first motions well, generally makes so good an impression, that every one is prepossessed in favour of what is to follow, and will rather excuse than condemn the little slips or mistakes that may be committed: whereas, if a bad impression is at first given, every little failing will be judged a crime, since prejudices of this kind take too strong a possession of the mind to be easily removed.

By the above direction it is presumed, that the General, who reviews, begins at the right, which they always do, unless the situation of the ground, or the drawing up of the Regiment, will not admit of it: which is a fault that should be carefully avoided. For this end, the ground on which you are to be reviewed, and the avenues leading to it, should be considered, and the most advantageous part of it pitched upon for the Regiment; taking care to draw up the front towards the place by which the General is to approach, and leaving the right flank open, that he may come to it without any difficulty. If this precaution is neglected, the Officer, who commands the Regiment, will be thought either careless or ignorant in his profession,

fession, unless it plainly appears that necessity, and not choice, obliged him to it.

When the case happens that the General comes to the left of the battalion first, and passes along the front to the right, the drummers are to cease beating when he comes to that flank, 'till the Regiment has faced to the right, as before directed, that the words of command for the facing may be distinctly heard.

When the battalion is ordered to face, the Officers, Serjeants, Drummers, and Musicians, are to do the same, and all to remain at their posts, without going through the battalion to the rear, when the General passes along it, or saluting him any more than once standing.

N. B. When the Officers pull off their hats, after saluting, they are not to bow their heads.

ARTICLE III.

As soon as the ceremony of viewing the regiment standing is over, the General then acquaints the Colonel what he would have performed, as the going through the Manual Exercise and the firings, or a part of each; all which depends entirely on the directions he shall be pleased to give, and therefore no certain rule can be prescribed. And though the Generals are not tied down to any set form in reviewing, yet they commonly proceed in the following manner.

First, They view the Regiment standing.

Secondly, They order the Manual Exercise and evolutions to be performed.

Thirdly, to go through some part of the firings; and,

Fourthly, To march by him, either in grand divisions, sub-divisions, or by single companies.

ARTICLE IV.

As the first part, that of viewing the regiment standing, has been fully treated of in the second article, I shall mention some things relating to the second (that of the exercise) which could not be so properly introduced before.

When the Officers are ordered to take their posts of exercise in the rear, the Colonel is not to go to the rear, but to march straight forward, and place himself by the General, with his esponton in his hand, during the exercise; and as soon as that is over, and the Officers ordered to the front, he is then to return to his post.

In the absence of the Colonel, the Lieutenant-Colonel is to proceed in the same manner, in placing himself by the General during the exercise, and performing all the other parts of the Colonel's duty in the command of the Regiment: but the Lieutenant-Colonel's post is never supplied by any other Officer when he is absent, or commands the regiment.

When the command falls to the Major, by the absence of the Colonel and Lieutenant-Colonel, he is then to take the Colonel's post at the head of the regiment, and salute with his esponton; but when the regiment is to perform the exercise, he is to mount on horseback to do it, the command of the regiment not being sufficient to excuse him from that part of his duty before a General, unless an impediment in his voice, or some other just reason, obliges him to decline it; and even in that case he is to make an apology to the General for his not doing of it himself, and desire leave that another officer may perform it; and when granted (which, I believe, is seldom refused) he then remains on foot, and acts in every respect as Colonel.

In the abfence of all the Field-Officers, the eldeft Captain takes the command, and places himfelf in the Colonel's poft, at the head of the regiment, and acts in every refpect as the Colonel fhould do, were he prefent.

When the Officers take their pofts in the rear, the Drummers and Mufick are to march directly forward, beating the *Troop*, till they have paffed the line, where the Major ftands, about fix paces; then facing to the right and left inwards, move on till they are directly behind the Major, the Mufick being in the center.

The Major, or Officer who gives the word of command, is not to find fault, or prefume to chaftife any of the foldiers in the General's prefence for any neglect in their exercife; neither ought it to be done before the Colonel, without his permiffion, in order to make them mind their duty, and inftruct them in the performance of it better; thefe things being only allowable at common exercife, and not in the prefence of our fuperior Officers: for which reafon, there fhould be nothing faid or heard but the words of command for what the men are to perform.

ACTICLE V.

After the Manual Exercife and evolutions, they perform the third part before-mentioned, that of the fireings; the directions for which, with the ufe and fervice of each fort, being fully treated of in the two following chapters, I fhall fay nothing further of it here; but proceed to the fourth and laft part of the ceremony of reviewing, that of marching by the General, either by grand-divifions, fub-divifions, or fingle companies.

When a battalion is divided into four equal parts or divifions, each divifion is then called a grand divifion.

Sub-divisions are formed by dividing each grand division into two equal parts.

By companies, is the marching of each company by itself, with its own Officers, serjeants and drummers.

The company of grenadiers is not included or told off in the grand or sub-divisions; but keep in a body by themselves on the right of the battalion, except when they are to fire, and then they are divided on the right and left.

When the regiment is to march off, the rear-ranks are to be closed to the front; which being done, the whole is ordered to wheel by grand or sub-divisions, or companies, as the reviewing General shall think proper.

If the battalion is to march off from the right, before they are ordered to wheel, the Colonel goes to the right, and posts himself at the head of the Captains on that wing; and the Lieutenant-Colonel is to post himself at the head of the Captains on the left.

When they are to march by grand divisions, and are wheeled to the right, they are to march in the following order.

I. The Major, advanced before the company of grenadiers, with their own Officers at their head.

II. The hatchet-men of the battalion, formed into ranks.

III. The Staff-Officers, *viz*. Chaplain, Adjutant, Quarter-Master, Surgeon and Mate.

IV. The musick, in a single rank.

V. The Colonel alone.

VI. All the Captains on the right, on the head of the first grand division.

VII. All the Lieutenants on the right, on the head of the first grand division.

VIII. All the Ensigns on the head of the third grand division.

Chap. V.   *Military Difcipline.*   69

IX. All the Lieutenants on the left, on the head of the rear, or fourth grand divifion.

X. All the Captains on the left, in the rear of the faid grand divifion.

XI. The Lieutenant-Colonel alone, in the rear of the Captains.

The ferjeants are to be divided equally to the grand divifions, and to march on the right and left flanks.

The drummers are to march upon the flanks of their refpective grand divifions, dreffing with the front rank.

The ranks being clofed forward to wheel, they are in marching to open to their former diftance of fix paces; for which end, the rear ranks are not to move till thofe in their front have got to their proper diftance, and then all the men in the next rank are to ftep at once forward with their left feet.

In marching, the Major is to falute on horfeback at the head of the grenadiers, being fome paces advanced before the Captain; but if he commands the regiment, he is then to march on foot in the Colonel's poft, and falute with his efponton.

The Officers are to march with their efponton downward, and when they come within twenty paces of the General, they are to bring them to their fhoulders, and to time their falute fo as to finifh and pull of their hats, a little before they come oppofite to him.

All the Officers, who march in the fame rank, are to be very exact in performing their motions together; and in order to have them done at the fame time, they are to be governed by, and to take them from the Officer who marches on the right of the rank they are in.

After the Officers have faluted, they are not to bow their heads in paffing by the General, but to march with their hats off till they have paffed him about eight paces, and then to put them on; and when they have got twenty paces from him, they are to bring their espontons from their shoulders, and march with them downward, as before.

The Ensigns are to carry the colours advanced, and to drop them when the other Ensigns drop the spear of their espontons (provided the General's rank entitles him to that compliment) and to march with the Colours down till they have passed him about six or eight paces. They are to pull off their hats when they drop the Colours, and not put them on till the other Ensigns do theirs.

The serjeants are to march with their halberts on their left shoulders, with the spears pointing downwards.

The grand divisions being subdivided as before directed; if the battalion is to march by sub-divisions, the Officers are to march as follows.

All the Captains on the right wing are to march at the head of the first and second sub-division.

The Lieutenants of that wing are to be divided on, and to lead the third and fourth sub-divisions.

The Ensigns are to be divided on, and to lead the fifth and sixth sub-divisions; the Ensigns with the Colours being posted at the head of the fifth.

The Lieutenants on the left wing are to be divided on, and to lead the seventh and eighth sub-divisions.

The Captains on the left are to march in the rear of the last sub-division.

The Field-Officers, Staff-Officers, Musick, and Hatchet-men, are to march in their former posts; and the serjeants are to be divided equally on the sub-divisions.

The drummers are to be divided to the sub-divisions, and march in the manner before directed for grand divisions.

The grenadiers are to march as before.

When they are to march thus, the battalion is to wheel by sub-divisions the Officers placing themselves

at

Chap. V.   *Military Discipline.*   71

at the head of the sub-divisions they are to lead, as soon as the ranks are closed forward, in order to wheel with them.

In marching, the ranks are to open to their former distance, as before directed; and where there are more Officers than one on a division, they are to observe the direction about the timing of their Salute, that they may do it together.

Reviewing by companies.

If the regiment is to march past the reviewing General by companies, the Major gives the word of command as follows:

To the right, close your files by companies!

At which the Officers, Serjeants, Drummers, and Hatchet-men, face towards their respective companies.

The right-hand file of every company stands fast, but all the rest face to the right.

When the Major perceives that the whole have faced properly, he gives the word,

March!

Upon which the Officers, Serjeants, Drummers, and Hatchet-men, step off, and march to their companies.

The companies step off at the same time, and close their files to the right.

This being done, he gives the word, *Halt!* upon which the whole face to their proper front. He then closes the rear ranks to the front, and wheels them to the right by companies, in the manner before directed for grand and sub-divisions. The order in which the companies are to march, is as follows.

I. The Captain.

II. The Lieutenant and Enfign in a rank, four paces in the rear of the Captain.

III. The ferjeants in a rank, four paces in the rear of the fubalterns, with their halberts advanced, in the fame manner as pikes were formerly.

IV. The drummers in a rank, four paces in the rear of the ferjeants.

V. The corporals and private foldiers, four in a rank; and if any odd men remain, they are to form the rear rank.

The Field-Officers are to march at the head of their own companies as Captains.

The Staff-Officers and mufick are always to march before the Colonel's company.

The hatchet-men may either fall into the ranks, or march before their captains, as the Colonel or commanding Officer fhall direct.

As foon as they have paffed by the General, either by grand or fub-divifions, or by fingle companies, they are to draw up on their former ground, unlefs ordered to the contrary, and to remain there till the General acquaints the Colonel, or Officer commanding the regiment, that he has no further commands for them; after which, the Colonel gives orders to lodge the colours, and difmifs the battalion.

ARTICLE VI.

When a regiment is ordered to march off from the left, either by grand or fub-divifions, the Colonel marches at the head of the Captains, who lead the left divifion, the Staff-Officers, Hatchet-men, and Mufick, marching before him, as formerly directed, when they marched off from the right.

The Lieutenant-Colonel goes to the right, and marches in the rear of the Captains on the right, who fall in the rear of the right divifion.

The

Chap. V.   *Military Discipline.*   73

The company of grenadiers marches in the rear of the Lieutenant-Colonel, and the Officers belonging to the company in the rear of it.

This is the method in all common marches, either in the line, or alone; but when the regiment is retiring from an enemy, or that any danger is apprehended in the rear, the Colonel remains there, and the Lieutenant-Colonel leads the battalion off.

CHAP. VI.

Consisting of directions for the different firings of the Foot.

ARTICLE I.

AFTER the exercise is over, the next thing to be performed is the different firings.

The battalion is supposed to be already divided into platoons; the number of which is usually sixteen, exclusive of the two grenadier platoons, but must depend upon the strength of the battalion, and the particular firings you intend to perform. Neither is a platoon composed of any fixed number of files; I mean those of three deep, which are commonly called half files, because a file of men are taken for six, in the ordinary way of speaking; whereas the true meaning of the word *file*, signifies all those men who stand in a direct line behind one another, or, in the military phrase, all those who stand in a direct line from front to rear: so that their being, three, four, or six deep, does not alter the sense, or change it from a file, but may be more or less, according as the battalion will allow of it: however, a platoon is seldom composed of less than ten files, which are thirty men, or more than sixteen files, which are forty-eight men: because, a platoon composed of less than ten files would not be of weight enough to do any considerable execution; and those above sixteen files would be too great a body of men for an Officer to manage upon service.

Chap. VI. *Military Discipline.*

In dividing the battalion into platoons, they should be composed of an equal number of files; or, at least, not above one file stronger than another; and those should be the flanks and colour platoons.

The Officers are to be posted as follows:

The Colonel, advanced before the reserve.

The Lieutenant-Colonel, in the front of the reserve.

The Major, in the rear of the first grand division, but to attend the right wing.

The Adjutant, in the rear of the fourth grand division, and to attend the left wing.

1st Captain, in the rear behind the reserve.

2d, to command the 1st platoon of the 1st firing.

3d, to command the 2d platoon of the 1st firing.

4th, to command the 3d platoon of the 3d firing.

5th, to command the 4th platoon of the 3d firing.

Captain-Lieutenant in the rear behind the interval of the 1st and 2d grand divisions.

1st Lieutenant in the rear, behind the interval of the 3d and 4th grand divisions.

2d Lieutenant to command the 3d platoon of the 1st firing.

3d, to command the 4th platoon of the 1st firing.

4th, to command the 5th platoon of the 2d firing.

5th, to command the 6th platoon of the 2d firing.

6th, to command the 1st platoon of the 2d firing.

7th, to command the 2d platoon of the 2d firing.

3d Ensign to command the 5th platoon of the 1st firing.

4th, to command the 6th platoon of the 1st firing.

5th, to command the 3d platoon of the 2d firing,

6th, to command the 4th platoon of the 2d firing.

7th, to command the 5th platoon of the 3d firing.

8th, to command the 6th platoon of the 3d firing.

The two eldest Ensigns are to carry the colours.

The Officers of grenadiers with their own men, as usual.

The serjeants are to be posted as follows.

The Serjeant-Major in the rear, chiefly about the center.

Sixteen in the rear rank, covering Officers that command platoons.

Two in the center of the center and rear ranks of the reserve, to complete the Lieutenant-Colonel's file.

Two in the center rank on the flanks, to complete the 2d and 3d Captains files.

Two in the center rank, to complete the 4th and 5th Captains files.

Four in the center rank, to complete the 2d, 3d, 4th and 5th Lieutenants files.

The serjeants of grenadiers remain with their own company.

The reasons for Officers being posted in the rear, are as follows; *First*, As the interval between each platoon should be but one pace, the Officer who commands the platoon is to fall into it when they fire; therefore, should any more Officers remain in the front, than one to each platoon, it would only embarrass and expose them to their own fire.

Secondly, It is of great use to have experienced Officers in the rear, to keep the men up, and see that they do their duty in action; as also to lead the battalion off in order when they are commanded to retire. And, *Lastly*, should there be no Officers in the rear when the battalion is ordered to the right-about, the men would be apt to march off too fast, and by that means break their ranks, and fall into confusion, or not halt in due time; which inconveniencies are prevented by Officers being posted there.

The method which is now practised, and which, by experience, is found to be most useful, is, the dividing of the platoons into three firings; each firing being composed of six platoons; which firings are not kept together in any one part of the battalion; but the platoons of each firing distributed, or disposed, into different parts of the regiment; the reasons for disposing of them into different parts, are these:

First

First, The difpofing of the platoons of each firing into different parts of the battalion, will extend your fire in fuch a manner, as to do execution in different parts of the oppofite regiment; the confequence of which, may either difable, or dif-hearten them fo much, as, upon a nearer approach, to oblige them to give way, or make but a faint refiftance.

Secondly, Their being divided in this manner, fhould the enemy and you join before thofe platoons have time to load, not any one part of your battalion is very much weakened by it: however, when the commanding Officer apprehends that this may be the cafe, he muft avoid it, by leaving off after the firft or fecond firings, that, they may be all loaded by the time they join the enemy, in order to throw in their whole upon them at once.

Thirdly, Should the platoons of each firing be together, too great a part of the battalion would be expofed in one place before the men could load, particularly the flank firings.

Fourthly, and *Laftly*, The firings being thus difpofed of, it makes the exercife appear more beautiful, and accuftoms the men to hear firing on their right and left, without touching their arms, till they have orders for it, which the *Englifh* are with difficulty brought to, from a natural defire and eagernefs to enter foon into action: a quality in fome cafes extremely commendable; but in others the contrary: for which reafon the men muft be taught to rely entirely on the conduct of their Officers, and to wait with patience for their orders, before they perform any motion; the due performance of which, both their fafety and honour depend on.

ARTICLE II.

For the better explaining, and the eafier comprehending, of the different firings, and the diftribution of the platoons

platoons of each firing in several parts of the regiment, as mentioned in the preceding article; I have hereunto annexed a plan of a battalion, told off, according to the present established practice, in eighteen platoons, eight sub-divisions, and four grand divisions, composing three firings of six platoons each.

The platoons marked with the letter A, are those of the first firing.

Those marked B, of the second firing.

Those marked C, of the third firing.

The letter D, marks the reserve.

By which means the different firings may be seen at one view, and how the platoons of each firing fall into the several parts of the battalion; and the whole appear so plain and easy, that, I believe, there will want no further explication for the comprehending of it.

The rule laid down in the plan for disposing the platoons of the different firings in the manner here mentioned, may be varied, if the commanding Officer thinks proper; because circumstances of time and place, or the situation of the enemy, may require a different disposition.

PLAN

Chap. VI.   *Military Discipline.*   79

PLAN of a battalion told off in eighteen platoons.

Grenadiers. C. 1st platoon of the 3d firing

A. 1st of the 1st } Right-hand sub-division
B. 1st of the 2d } and 1st to fire.   } 1st grand division to fire.

A. 3d of the 1st } Left-hand sub-division,
B. 3d of the 2d } and 5th to fire.

C. 3d of the 3d } Right-hand sub-division,
A. 5th of the 1st } and 3d to fire.   } 3d grand division to fire.

B. 5th of the 2d } Left-hand sub-division,
C. 5th of the 3d } and 7th to fire.

Colours. D. Reserve.

C. 6th of the 3d } Right-hand sub-division,
B. 6th of the 2d } and 4th to fire.   } 2d grand division to fire.

A. 6th of the 1st } Left-hand sub-division
C. 4th of the 3d } and 8th to fire.

B. 4th of the 2d } Right-hand sub-division,
A. 4th of the 1st } and 2d to fire.   } 4th grand division to fire.

B. 2d of the 2d } Left-hand sub-division,
A. 2d of the 1st } and 6th to fire.

Grenadiers. C. 2d of the 3d.

Form of the Battalion.

Previous

Previous to the performance of the firings, the Major is to prepare the battalion according to the following directions:

He first gives the caution,

> *Take care to prepare for the charge!*

Upon which all the Officers and serjeants are to advance their arms together at two motions. He then proceeds and gives the word,

> *Grenadiers, cover the flanks!*

Upon which the grenadiers face to the left, and the battalion to the right, in two motions, as above directed; the left hand file of the battalion excepted, which stands fast. The hatchet-men face to the right-about, and march without any further order, two paces beyond the rear-rank, facing again immediately to the right, towards the center of the battalion.

> *March!*

The drummers beat the *Point of War*, accompanied with the Fifers: the division of grenadiers for the left, consisting of one half of the company, commanded by the First Lieutenant, marches briskly in the rear of each rank, to the left flank: the right hand platoon of the battalion, stepping off with the left feet, marches twenty-nine small paces to the right: the next platoon steps off with the left feet, when the first makes its third step, and marches twenty-seven small paces: the 3d platoon on the right, marches off with the left feet, when the preceding one makes its third step, and goes twenty-five paces to the right; and so on, as in opening the files. The grenadiers of the right

right, move on slowly, till they come within three paces of the ground, where the right of the Battalion halts: by this means, there will be an interval of one small space betwixt every platoon, through which the supernumerary Officers are to go to the rear. The Officers, who are to command platoons, place themselves opposite to their respective intervals, as do the Serjeants in the rear, the Hatchet-men march, till they are exactly in the rear of the Colours. The Drummers, being divided into four divisions, are to take post behind the center of each grand division; the two orderly Drummers excepted, who are to remain in the front, with the commanding Officer. When the second platoon on the left, has stepped off with the left feet, the Major gives the word, to *Halt*, as the platoon, on the left of all, is not to move. On the word *Halt!* the Battalion faces to the left, and the Grenadiers to the right, the Officers and Serjeants ordering their arms at the same time. The Drums and Fife are also to cease; and the Major proceeds.

Fix your bayonets! 10 motions.

As in explanations, the tenth, eleventh, twelfth, and thirteenth, of the manual exercise.

Shoulder! 5 motions.

As in explanations the fourteenth and fifteenth.

Prime and load! 21 motions.

See explanation the twenty-ninth.

Rear rank close to the front!

Upon which the Officers on the right of the Colours, go to the right-about; and those on the left of the Colours, to the left about, in three motions, at the first and third of which, they are to advance their arms, the Serjeants taking care to do the same.

March !   10 motions.

The rear ranks close, and the Officers go into the intervals of the platoons, as the Serjeants do from the rear : the supernumerary Officers remain four paces in the rear.

The Major then acquaints the Colonel, that the Battalion is ready, and rides to his post, in the rear of the right. The Colonel advances his Esponton, and having received the General's orders for the firings, &c. takes his post at the head of the Reserve, and then proceeds to fire the Battalion.

N. B. In the firing by grand divisions, the Officers must give a little more time betwixt each fire, that one half of the Battalion may always be loaded.

Before I proceed further, it will be necessary to explain the platoon exercise ; that is, what number of motions of the manual exercise they are to perform at each word of command.

There are but three words of command used in the platoon exercise, which are as follows.

I. *Make ready !*   II. *Present !*   III. *Fire !*

By the first word of command, the men are to join their right hands, and recover their firelocks in two motions, as directed in explanation the first of the manual exercise, placing the thumb upon the cock, and the fore-finger behind the guard. Then tell 1, 2, and cock their firelocks ; in doing which, the front rank steps back with the right foot in a direct line to the rear, and kneels upon the right knee, in the manner described in explanation the thirtieth.

The center rank steps back with the right feet eighteen inches, as in explanation the thirty-third ; and the rear rank

Chap. VI. *Military Discipline.* 83

rank fteps with the right feet, to the right, as in explanation the thirty-fixth.

The placing the feet in this manner, is called, in military terms, *Locking*.

Formerly the men in each file ftood in a direct line
behind one another, by which means thofe in the center
rank were obliged to ftoop, that the men in the rear
rank might fire over their heads; but by the above pofition, that inconvenient and uneafy pofture is avoided:
for by locking as aforefaid, the men of the center
rank prefent their firelocks over the right fhoulders of
their file leaders; and thofe of the rear rank prefent
to the right of the center rank men, which brings their
firelocks, when they prefent, to the intervals between
the files.

By the fecond word of command, they are to prefent their firelocks, as is directed for the feveral ranks,
in explanations the thirty-firft, thirty-fourth, and thirty-
feventh.

By the third word of command, they are to fire, as
in explanation the thirty-fecond, after which they are
to proceed to half-cock, prime, load, and fhoulder,
performing the feveral motions together, as directed in
the exercife.

In dividing the platoons into the feveral firings,
particular care fhould be taken, to let the Officers and
Soldiers know diftinctly what firing they belong to;
whether of the firft, fecond, or third firing; as alfo
what number each platoon is of in the different firings,
as the firft, fecond, third, fourth, fifth, or fixth platoon, of the firft, fecond, or third firing; that no miftake may happen in the execution.

In order to know if they have rightly underftood
their tellings off, the Major, or Adjutant, may try
the platoons of each firing apart, by making thofe
platoons perform fome motions together, as poifing
and fhouldering; the fame may be done by the platoons of each firing fingly. This, however, fhould

G 2 be

be practised only at common exercise, or before the General, who is to see you go through your firings, comes into the field.

At the beating of a *Preparative*, all the platoons of that firing, which is to come next, are to make ready together, as explained in the platoon exercise.

A *Flam*, or double stroke, is the signal for the platoons to begin to fire; the particular directions for which shall be treated of in the following article.

Whenever the Battalion is to advance, or retreat, the commanding Officer is to make his orderly Drummers beat the first division of the *March*, or of the *Retreat*, and immediately upon their beginning it again, the Battalion is to move.

The Battalion is always to halt when the Drum ceases; in order to mark which the better to the men, the orderly Drummers are to finish their beats with a strong *Flam*.

ARTICLE III.

The Major having prepared the Battalion for the charge, according to the foregoing directions, and taken his post in the rear of the right, the commanding Officer is to apprise the Battalion, whether he would have them begin by platoons, by sub-divisions, by the firings, or by grand-divisions, before he orders the *Preparative* to be beat: and whenever he would make any change in the firings, he is to order the first part of the *General* to beat, to cease firing; and then acquaint the Battalion, what other firing shall follow.

To go through the firings standing.

If the Battalion is to begin, in the usual way, by platoon *standing*, the commanding Officer is to order his Drummers to beat a *Preparative*, at which the whole

Chap. VI.   *Military Discipline.*   85

whole Battalion is to make ready, the front rank coming down together, as is directed in the platoon-exercise.

In all firings, the Officers and Serjeants, who are posted to platoons, are to keep in the intervals of them; and to carry their Espontons, and Halberts, advanced to their right sides.

The whole Battalion having made ready at the *Preparative*, as above, the Officer commanding the first platoon at the first firing, begins, and gives the following words of command.

Present !   Fire !

As soon as the word, *Fire*, is given to the first platoon, the Officer commanding the second platoon is to give the word *Present!* and then, *Fire!* The Officer commanding the third platoon, is to observe the same rule when the word *Fire*, is given to the second platoon: all the other Officers commanding platoons are to follow the same directions.

The Officers, who give the words of command, are to speak them clear and distinct, and not to proceed to the word, *Fire*, till the men have presented as they ought, or wait too long after they have.

When the words of command are given with judgment, the fire is generally good; so that the firing we or ill, depends, in a great measure, on the manner th Officers give them.

Before the platoons of Grenadiers kneel and lock, they are to wheel to the right and left, inward, an eighth part of the circle, which, as being on the flanks, they are always to do, in whatever firing they are placed.

By wheeling the Grenadiers inward, in this manner, it throws their fire towards the center of the opposite Regiment, and consequently rakes a great part of their front, which, if they level well, cannot fail of doing confiderable execution, fince every ball muft have its effect.

As the platoon exercife directs the men to load, as foon as they have fired; and, when loaded, to fhoulder their firelocks, I thought it unneceffary to mention it at the end of each firing.

The fignal for ceafing firing, is the firft part of the *General*, as has been before obferved; and is ufually given, when the three firings have finifhed the fecond round: however, that entirely depends upon the commanding Officer, and the firing is to continue, till he thinks proper to make the fignal for its ceafing.

The Battalion having gone through their firings by platoons, ftanding, I fhall, in the next place, give directions how they are to perform them advancing.

How to fire by platoons, advancing.

Upon the commanding Officer's ordering the Drummers to beat a *March*, the whole Battalion is to march ftraight forward, beginning with their left feet, and to move as flow as foot can fall.

When the Battalion has advanced a little way, the *Preparative* is to beat, at which the platoons of the firft firing make ready, the two firft ftepping brifkly forward three good paces, and the front ranks coming down at the third.

The Officer commanding the firft platoon then begins, and gives the word, to *prefent* and *fire*, as before.

When the firft platoon fires, the third is to lock.

When the fecond fires, the fourth is to lock.

When the third fires, the fifth is to lock; and fo on through the whole Battalion, every Officer taking care to have at leaft an interval of one platoon between

Chap. VI. *Military Discipline.* 87

tween his firing, and that of the platoon next but one before him.

When the fifth platoon of the first firing, fires, the second firing makes ready, the first platoon of which, at the same time, advances and locks; and is followed by the remaining five platoons of that firing, in the manner directed for those of the first firing.

The same is to be observed by the third firing, when the fifth of the second fires.

When the fifth platoon of the third firing has fired, the first of the first firing locks again, in order to begin the second round; and so on; the fire being continued, till the commanding Officer orders the first part of the *General* to be beat, at which it is immediately to cease; but the Battalion is still to keep advancing, as long as the *March* is beat, the ceasing of which, is the signal for it to halt, as has before been observed.

How to fire by platoons, retreating.

The Battalion having thus gone through the firings by platoons *advancing*, the commanding Officer orders his Drummers to beat a *Retreat,* at which the whole goes to the right-about, and marches with a very slow pace to the rear.

The Battalion having marched a little way, the commanding Officer orders a *Preparative* to be beat; upon which the six platoons of the first firing make ready, and the two first to fire, come to the right-about, and lock.

As soon as the first platoon has fired, the third comes about, *&c.* when the fourth platoon has fired, the second firing makes ready; and as soon as the fifth platoon fires, the first of the second firing comes about, and so on, as before.

When the Battalion has thus fired *retreating,* as many rounds, as the commanding Officer shall think proper,

proper, he orders the first part of the *General* to be beat; and when the Drum ceases beating the *Retreat*, the Battalion halts, waiting in that position for a *Flam*, which is a signal for the whole to come to the right-about again to their proper front.

How to fire by sub-divisions, standing.

At the *Preparative* for firing by sub-divisions, the right-hand sub-division of each grand-division makes ready.

When the right-hand sub-division of the first grand division has fired, the right-hand sub-division of the fourth grand-division follows: Then that of the second followed by that of the the third.

When the third sub-division has fired, the four left-hand sub-divisions make ready, and proceed as the first four did.

When the Battalion has fired twice, the commanding Officer fires the two platoons of Grenadiers, by word of command from himself, which are to move out, and lock, when the seventh sub-division fires its second round.

How to fire by sub-divisions, advancing.

The orderly Drummers being ordered to beat a *March*, the Battalion steps off with the left feet.

At the *Preparative*, the four right-hand sub-divisions make ready, the first of which to fire, springs forwards three paces, and locks. On the word *Present!* the second, or right-hand sub-division of the fourth grand-division, advances three paces, and locks; and so on.

The Grenadiers fire last, as before, by word of command from the exercising Officer.

How to fire by sub-divisions, retreating.

At the beating of the *Retreat*, the Battalion goes to the right-about, and retires flowly to the rear.

At the *Preparative*, the right-hand sub-divisions make ready, and the first comes about, and locks. On the word, *Present!* the second to fire comes about, *&c.* When the third sub-division fires, the four left-hand sub-divisions make ready; and when the fourth of the right-hand sub-divisions *presents,* the first of the left-hand sub-divisions comes about, and locks, *&c.* The Grenadiers fire together, as before, by word of command from the front.

How to fire by grand-divisions, standing, advancing, and retreating.

This is performed in the manner directed for the sub-division firing.

How to fire by the firings, standing.

At the *Preparative*, the six platoons, which compose the first firing, make ready; but the front rank is to wait for a *Flam* to kneel.

The commanding Officer gives the word to the whole, to *present* and *fire* together.

As soon as the first firing has loaded and shouldered again, a second *Preparative* is beat, at which the second firing makes ready; the front rank waiting for a *Flam* to come down, as before directed for the first.

The like is to be observed by the third firing, the Grenadiers making ready at the same time, as they compose a part of that firing.

How to fire by the firings, advancing.

The Battalion being advancing, the commanding Officer orders a *Preparative*, upon which the whole is to halt, and the six platoons of the first firing to make ready; the front rank waiting, as before, for a *Flam* to come down.

The commanding Officer, having given them the word, to *fire*, immediately orders the *March* to be beat again, upon which the Battalion advances, those platoons that fired, loading as they march.

At the next *Preparative*, the whole Battalion halts again, and the second firing makes ready. After it has fired, the *March* is beat, and the Battalion moves forward again; the platoons of the second firing, loading as they march, in the same manner, as directed for those of the first.

The same is to be observed, with regard to the third firing.

How to fire by the firings, retreating.

Upon beating the *Retreat*, the Battalion goes to the right-about, and marches off towards the rear. At the *Preparative*, the Battalion halts, and the first firing makes ready, there waiting for a *Flam*, at which the whole comes to the right-about, and the firing locks. The commanding Officer having given them the word to *fire*, orders the *Retreat* again, at which the Battalion goes to the right-about, and moves off, the platoons that fired, loading as they retire.

The same is to be observed, with respect to the second and third firing.

When foot are attacked by horse, whether they are drawn up in line of battle, or in a square, it is
proper

proper for them to reserve their whole front rank, and fire only their center and rear ranks by platoons. In this case, the front rank is not to fire, till they are put to the last extremity; and not even then, till the horse are so close, that they have but just time to charge their bayonets, after they have fired; that they may be sure of placing in them, both their ball, and bayonets, in almost the same instant of time. But I shall treat of this more at large hereafter.

ARTICLE IV.

Firing by ranks.

To fire by ranks, is meant, to fire only one rank of the Battalion at a time, beginning first with the rear rank, then the center rank, and, lastly, the front rank. The manner of performing it is as follows:

The whole Battalion is to make ready at the same time, and immediately kneel and lock, as in the platoon exercise. Then the commanding Officer gives the following words of command.

Rear rank, Present! Fire!

As soon as the rear rank has fired, they are to recover their arms, fall back to their former distance, prime, load, and shoulder. After the rear rank has fired, the Major proceeds, *Center rank, Present! Fire!* After firing, the center rank recovers their arms, falls back, primes, loads, and shoulders. When the center rank has fired, the front rank is to do the same; which may be done either kneeling, or by making them stand up first.

In the time that pikes were in use, I presume that this was the method prescribed when attacked by horse, the whole front rank being composed of Pike-Men, and the center and rear Musketeers. When
the

the Musketeers were ordered to make ready, I suppose the Pike-men kneeled down, as the front rank does now, dropping the spears on the ground till the two ranks of Musketeers had fired, and then rose up and charged their pikes, remaining in that position till the Musketeers had loaded.

As I never had any experience with the pikes, they being laid aside just when I came into the service, I hope I may be excused, if what I have here mentioned is wrong; but as the firing by ranks, both in the Battalion and the square, was practised a considerable time after the pikes were gone, I presume, from thence, that it was their method, and retained by the old Officers, who laid a great stress upon it, as the most effectual way to secure them against horse. But this is not to be wondered at, since it is natural for all mankind to be prejudiced in favour of the first notions they receive, or customs which they have been long used to: however, it is seldom or never used in service, though sometimes practised in the exercise; but another method is substituted in its room; which is, that of saving the fire of the whole front rank of the Battalion to the last, and firing the two rear ranks by platoons; it being the compact fire which does the execution requisite to break a Squadron; whereas the fire of a single rank is so thin, that it will not easily stop their progress, if their resolution do not fail them.

ACTICLE V.

Parapet firing.

This firing is only used in fortified towns when besieged, in intrenchments that are attacked, or that you are to fire over a hedge, or wall, at the enemy.

There are two ways of performing it; the one by ranks, and the other by files.

By ranks.

As the breaft-work, parapet, or hedge, is before the men, they are obliged to fire ftanding, and therefore no more than one rank can fire at a time, which begins with the front rank, who, as foon as they have fired, are to form in the rear, that the center rank may march up and fire; and when they have fired, they are to form in the rear alfo, that the rear rank may march up and do the fame.

There are two ways of performing it by ranks; the one with the files open, and the other with them clofed.

When the files are open, as foon as the front rank has fired, they are to recover their arms, face to the right-about, which brings them directly oppofite to the intervals on their right, through which they are to march to the rear, and then to face to their proper front, by going to the right-about, which brings them in the rear of their own files, forming then the rear, of which before they were the front. Upon the front rank's marching down the intervals to the rear, the center and rear ranks march forward, the center into the ground from whence the front rank fired, and the rear rank into that where the center ftood. When the center rank comes into the ground of the firft, they are to fire, recover their arms, face to the right-about, march to the rear, and face again to their proper front. The rear rank is to do the fame; fo one rank after another, as long as the commanding Officer fhall think proper. By this means you may keep almoft a conftant fire, fince the time between each will be very inconfiderable.

This manner of performing it, with the files open, is much in the fame nature as counter-marching by files.

Before

Before they begin to fire, the commanding Officer is to order the whole Battalion to make ready, proceeding no farther than recovered arms, till the signal is given for them to fire. As soon as the ranks that have fired are formed in the rear, they are immediately to prime, load, and make ready; and march forward with recovered arms, as the ranks before them move up to fire without any further word of command than that which was first given: which directions will serve for all parapet firing.

How to perform it by ranks, with the files closed.

In this case, the Battalion must be told off by platoons, leaving an interval of a large pace between each. When the whole front rank of the Battalion has fired, which it is to do in the same manner as that with the files open, the men of that rank are to recover their arms, and face to the left; but the left-hand man of each platoon must face to the left-about, which brings him opposite to the interval on the left of his platoon. As soon as they have faced, the left-hand men of the platoons are to march straight down their several intervals to the rear, all the rest following them to the left; and as each man comes opposite to the interval on the left of his platoon, he is to face again to the left, and march down the interval, those of each platoon following their left-hand man. As soon as the left-hand man of each platoon has got one pace beyond the rear rank, they are to face to the left, and continue marching till they come to the right of their own platoons, the others following in file in the same manner, and then halt, forming then the rear-rank, of which before they were the front. After they are formed in the rear, they are immediately to load, and, as soon as loaded, to make ready, and march forward, as before directed.

When the front rank has fired and marched clear of the front, the center and rear ranks are to march forward, the center into the ground of the firſt, and the rear into that of the center; then the center rank is to fire, recover their arms, face to the left, march down the intervals, and form in the rear of their own platoons, as the front rank did. The rear rank is then to march forward, fire, and form in the rear, as the others.

This is a ſort of a counter-marching by ranks, by bringing the left of each platoon to the right, and the right to the left; with this difference, that, inſtead of keeping the ſame ſtation, each rank in its turn forms the rear, and as thoſe before them fire, they move up to their former ground.

I think I have no occaſion to give my opinion which of the two ways, that with the files open, or the other with them cloſed, is the beſt, ſince every body will agree, that the one which contains the moſt fire, which is that with the files cloſed, muſt have the preference.

Parapet firing by files.

The Battalion muſt be told off into platoons, as in the other firing, and drawn up at three paces diſtance from the parapet, breaſt-work, or hedge, and the whole ordered to make ready together as far as recovered arms: and when the ſignal is given for them to begin to fire, the files on the right and left of each platoon (that is, one file from the right, and one file from the left, of each platoon) move forward, and when the file-leaders come up to the breaſt-work, or hedge, they are to face to the right and left inwards (that is, the right and left hand men of each platoon facing towards one another) thoſe of the center and rear following their file leaders, till the two front men join, and then they are to halt: by which the two files of each platoon form a rank of six men in the front of their platoons. As soon as they are thus formed in the front, they are to present and fire ; then recover their arms, face to the right and left outwards, and march back to their own places in the manner they came. When the first files have fired, those files which stood next them are to march out, and draw up in the front of their platoons, in the same manner as the others did, and fire; then recover their arms, and march back to their former places. The two next files of each platoon are to march out, fire, and return to their places in the same order as the others, and so on till the two center files have fired ; after which the flank files of the platoons are to begin again, unless ordered to the contrary.

To avoid confusion in their drawing up in the front of their platoons to fire, the file-leaders should always form in the center, as before directed, those men of the center and rear ranks drawing up on the outside of them ; which must be done when the two center files move out, they being to march up straight to the parapet, and the men of the center and rear-ranks to face outward, and draw up on the right and left of their file-leaders: however, it may be done otherwise with the two center files of each platoon, by making the file-leaders face to right and left outwards when they come up to the parapet, bringing the men in the rear to form in the center of each rank.

As soon as the files are returned to their places, after firing, they are to face to their proper front, prime, load, make ready, and wait with recovered arms to fire again in their turn : all which must be done without any further word of command than what was at first given to begin ; and not to discontinue it, till ordered so to do.

ARTICLE VI. *Street-firing.*

It is so called from your being obliged to engage in a street, high-way, lane, or narrow passage, where no more than ten, twelve, sixteen, or twenty files, can march in front; so that according to the breadth of the place, your platoons must be stronger or weaker.

The manner of performing it at exercise, is thus:

The rear-ranks are to be closed forward to close order, and the Battalion is to wheel to the right or left by double platoons, or sub divisions.

By the wheeling of the sub-divisions, they fall in the rear of one another; so that no more than one sub-division can fire at a time.

As soon as the sub-divisions have wheeled, they should march in that position, because, the firing will appear more graceful when it is begun while the Regiment is in motion, than when it stands still. For as that which is performed in motion, carries a greater resemblance of real service than the other, it must therefore, by so lively a representation of action, raise the imagination to a higher pitch.

When the whole Battalion is in motion, the commanding Officer should give the signal for the firing to begin; on which, the Officer who commands the front sub-division, is to order his men to make ready, present, and fire; and as soon as they have fired, they are to recover their arms, face from the the center to the right and left outwards, march down the flanks of the other platoons, and form again in the rear of the last.

The proper time for them to begin their loading motions, will be about as soon as they have passed the third sub-division. When the first sub-division presents, the Officer who commands the second, must
order

order his men to make ready, and march up with recovered arms to the ground the firſt fired on, as ſoon as the others have got on the flanks; and when his men have kneeled and locked, he is to give the Words, *Preſent! Fire!* and when fired, to recover, face outwards, march along the flanks, and form in the rear of the firſt. The reſt of the ſub-diviſions are to obſerve the ſame directions in making ready, marching up to the ground on which the firſt fired; and when fired, to march and form in the rear.

The ſub-diviſions are to keep up pretty cloſe to one another, and to move or halt as thoſe in the front do.

When this is to be put in practice on real ſervice, the front of the ſub-diviſions muſt not be equal to the breadth of the place you are to engage in; but there muſt be a ſmall ſpace of ground, or interval, left on your flanks, that thoſe who have fired may have room to march back and form in the rear.

It is in this manner, when you have not time to raiſe a breaſt-work, that a paſs, bridge, road, or ſtreet, is to be maintained againſt the enemy, by the ſub-diviſions ſuſtaining one another and firing in their turn; which may be continued as long as there is an occaſion, almoſt without intermiſſion by one Battalion only.

ARTICLE VII. *Running-fire.*

This fire is never made uſe of but upon the gaining of a battle, the taking of a town, the celebration of the King's birth-day, or thoſe of the Royal Family, or ſome other extraordinary cauſe of rejoicing; for which reaſon the *French* call it a *Feu de joye*.

Theſe firings are always performed in the duſk of the evening, both in camp and garriſon. The ranks are to be cloſed no nearer than half diſtance, the front rank being to ſtand as well as the center and rear; and, when

when they present, they are all to raise their muzzles pretty high, in order to fire in the air. The men of each file are to fire together; that is, each file distinctly by itself; and so run pretty quick from one file to another, quite through the Regiment.

The manner of performing it in camp.

As soon as the Sun sets, the Army is to draw out at the head of their encampment; or if the ground will allow of it, both the lines may be drawn up in the front of the first line of tents. The train of artillery is likewise drawn out on these occasions, and placed at the head of the first line, or upon a rising ground, if any such lies near them.

The firing is to begin with the train, keeping such time between each gun that twenty-five or thirty may be fired in a minute.

It is a fixed rule to fire an odd gun, as 21, 31, &c.

As soon as the train have fired the number of guns appointed them, the fire of the small arms is to begin on the right of the first line, running gradually on from file to file, and from Regiment to Regiment, till it comes to the left of the first line; then it is to begin on the left of the second line; and run on gradually in the same manner to the right of that line, which finishes the first fire of the whole army; after which they are all to give three huzzas, then load and shoulder.

As they are to fire three times on these occasions, the other two are to be performed in the same manner as the first, beginning with the Artillery, from thence with the right of the first line, and ending with the right of the second line, giving three huzzas after each fire is quite ended.

To prevent the fire running too quick, the Regiments in the first line should not make ready till that on their right has begun to fire; and those in the second line not to make ready till the Regiment on their left begins to fire; those in the first line being to take it from the right, and the second line from the left: for should they all make ready together, the center or left would be apt to fire as soon as it began on the right; but their not making ready till the Regiment, from whom they are to take it, begins to fire, will prevent their firing too soon, which fault is usually committed in these firings; but very seldom that of being too slow.

The manner of performing it in Garrison.

The Garrison is to be drawn up on the ramparts, extending themselves quite round the town, if their numbers will allow of it, and to face the parapet, over which they are to fire.

The artillery, as in camp, is to fire first; then the small arms, beginning on the right of the eldest Regiment, and to run gradually round to the left. After the fire ceases, the whole Garrison is to give three huzzas, then load and shoulder. The other two fires are to be performed in the same manner; as also the above directions about the time of making ready will serve likewise in garrison.

Having gone through the different firings, as proposed, I shall give some directions, in the following Chapter, how foot are to proceed when attacked by horse, both in Battalion and in the square, but that I may keep within due bounds, I shall confine myself to the management of a single Battalion.

CHAP.

CHAP. VII.

Containing directions how a Battalion of foot is to defend itself when attacked by horse..

ARTICLE I.

AS foot are sometimes interlined with horse, or detached from the main body to secure some important post, by which they are exposed to the attacks of horse, it will be proper to lay down some general rule how a Battalion is to proceed on such an occasion, both as to the management of their fire in Battalion, when only attacked in front; and in what manner they are to throw themselves into a square, when their flanks and rear lie open and exposed, and how they are to fire and march when formed in the square.

When a Regiment is to march through a country, or posted at a place, where there is a possibility of their being attacked by horse, they should be prepared to defend themselves against them, by dividing their platoons in such a manner that they may have a constant succession of fire, when only attacked in front; or be ready to form the square when necessary, without any new telling off.

If the Battalion is strong enough to admit of it, I would recommend the plan in the foregoing chapter, consisting of sixteen platoons besides Grenadiers, which composes three firings of six platoons each: and, if the commanding Officer thinks proper, the front rank of every firing may be kept, as a reserve. But the chief point is, that the square is safer, easier, and quicker formed from this plan, than any other now in use, as will appear by the said plan, when I come to treat on the forming of the square. But lest the Battalion should not be strong enough to admit of sixteen platoons, I have annexed another of twelve platoons besides the Grenadiers, from which the square may be formed in the same manner as the first. It will likewise consist of three firings, of four platoons each, besides the Grenadiers; in which case, it will be very proper to keep the whole front rank and the Grenadiers for the reserve.

Plan of a Battalion told off in twelve platoons.

Grenadiers, D. Reserve.

> A. 1st platoon of the 1st firing.
> B. 1st platoon of the 2d firing.
> C. 1st platoon of the 3d firing.
> A. 3d platoon of the 1st firing.
> B. 3d platoon of the 2d firing.
> C. 3d platoon of the 3d firing.

Colours. D. Reserve.

> C. 4th platoon of the 3d firing.
> B. 4th platoon of the 2d firing.
> A. 4th platoon of the 1st firing.
> C. 2d platoon of the 3d firing.
> B. 2d platoon of the 2d firing.
> A. 2d platoon of the 1st firing.

Grenadiers. D. Reserve.

Front of the Battalion

If foot could be brought to know their own strength, the danger which they apprehend from horse would soon vanish; since the fire of one platoon, given in due time, is sufficient to break any squadron: therefore, if a Battalion

talion of foot would manage their fire to the beft advantage, and not throw it away at two great a diftance, which they are apt to do, from their appearing nearer than they really are, by their being fo much above the foot, they might baffle a confiderable body of horfe, and make them defift in a very fhort time from any further attempts upon them.

But as the horfe will have recourfe to ftratagem to draw away your fire, by making feint attacks, with fmall parties advanced before the body, in hopes to make you fpend your fire on them; the commanding Officer, however, may, without any hazard, eafily difappoint their defigns in the following manner:

Let us fuppofe a Battalion drawn up where the horfe can only attack them in front, the flanks and rear being fecured by moraffes, rivers, hedges, or ditches. In fuch a fituation, one Battalion of well-difciplined foot may defpife the attacks of a whole line of horfe, while they continue their attacks on horfeback, and oblige them to retire with confiderable lofs.

We will fuppofe then a Battalion pofted as above, and a body of horfe, having no other way to pafs than through that which is occupied by the foot, obliged to attack them in that fituation.

In this cafe, the Officer who commands the cavalry will, no doubt, form them into feveral lines, in order to fuftain one another, not doubting but the firft and fecond lines will be forced to give way by the fire of the foot; and in all probability they may be ordered to advance with no other view then to receive the fire, and then retire through the intervals of the fquadrons, which are marching to fuftain them; imagining that two or three feint attacks of this kind will be fufficient to draw away all their fire, and give the reft an opportunity to fall upon them before they can have time to load again: but if the fire of a Battalion is managed according to the directions of my firft plan, which is divided into three diftinct firings, they can never be without one or more fires, for every attack they can make: for if the lines of horse do not leave a considerable distance between each, they will run a great hazard of being broke, and thrown into confusion by their own troops, who are ordered, or obliged to retire; which the three first attacks, with any tolerable conduct in the Officer who commands the Battalion, will certainly be obliged to; and if they leave proper intervals between the lines of the horse, it will give the foot time, notwithstanding the quick motions of the cavalry, to load, or at least very near it, before they will have an occasion to make use of a second fire. But let them attack after one another as quick as the nature of the thing will admit of, the platoons of the first firing will be loaded before they can possibly have an occasion to make use of those of the third firing; so that the Battalion can never be without two firings; for which reason I do not think there is an occasion to reserve the whole front rank, which addition of fire to each platoon is of great consequence, and, in my opinion, of infinite more service than it can be of when reserved to the last; particularly so, since there is a great probability that you will not be reduced to the last fire; and if you are not reduced to the last fire, the front rank is rendered useless by reserving it, the fire of which might do considerable execution in firing along with their platoons. However, the commanding Officer will see, by the disposition of the enemy, whether it is necessary to reserve the front rank, or not. His own reason must direct him in that affair, the rules laid down here being rather general than positive; the variety of circumstances which happen in action rendering it impossible to determine absolutely on this head.

I shall now return to my former proposition, that of disappointing their designs, in drawing away your fire by feint attacks.

An Officer, who has had any experience, may discover the designs of the enemy by the disposition of their troops; particularly in the case we now suppose. If you find them formed into several lines, you may conclude it is to make several attacks immediately after one another, and that the first and second are only designed as feints to draw away your fire; for which reason it would be proper to order three or four small detachments, of four or five files each, taken from different parts of the Battalion, to advance ten or twelve paces in the front, and when the horse comes within thirty or forty paces of them to fire, and then retire immediately into their places. If this was only designed as a feint, they will retire at that fire; but if it was not a feint, though it may not be sufficient to break them entirely, yet it may do them considerable damage, and put them into some disorder, particularly if any of the Officers should be killed or wounded. If those squadrons should advance after that fire, they must be received by the platoons of the first firing, which, I am convinced, will send them back faster than they came on, unless their horses are ungovernable, and by that means bring some of them forward contrary to their inclinations.

The detachments, or small platoons, so advanced, should be taken out of the platoons of the third firing, by which they will have time to load, after they return, before there will be an occasion to make use of that firing.

If the squadrons of the first line retire at the fire of the advanced platoons, in order to make room for the second line to advance, you may serve them in the same manner, by advancing the same number of small platoons out of the same firing, there being no fear from the want of time, since the second line cannot charge till the first have got clear of their front. Besides, if the second line is too near the first, they
will

will be in great danger of being broke by them; to avoid which, they will leave proper intervals between the lines, as well as between the squadrons; so that you cannot fail of time to put it in execution, provided proper care was taken beforehand to make the disposition proposed.

By this disposition, every attack will receive two fires, after which, I believe, there is no great danger of their advancing; but if they should, the platoons of the second firing are ready to be made use of.

Some may object against the advancing of the little parties, as not being sufficient to break or repulse the squadrons, and therefore give them an opportunity to charge those parties before they can join the Battalion; but as they are only small platoons, and advanced but a very little way from the front, they can fall into their places after they have fired in a moment, and consequently avoid the danger with a great deal of ease.

Those parties should not advance before the Battalion, till the cavalry are in full march to attack you; lest they should discover your design, and order their attacks accordingly.

When the advanced parties make ready, the platoons of the first firing should do the same; but great care must be taken that they do not fire till the advanced platoons are returned, and even not then till the horse are within twenty-five or thirty paces: for on the keeping of your fire depends your safety. If on presenting, the horse should make a full halt, or wheel off, as they frequently do, the men must be cautioned not to fire, but immediately recover their arms without firing, lest they should do that only by way of feint to draw away your fire at some distance, and then make a real attack, hoping to find you unprovided to receive them.

When foot are once brought to that perfection of discipline, as to recover their arms, after they are

presented, without firing, in the face of the enemy, the horse will never pretend to attack them a second time, but keep their due distance: but if they throw away their fire too soon, they will take the advantage of it, and be upon them in an instant; and if they can once penetrate but with one squadron, it will throw a Battalion of six hundred men into confusion; after which their conquest will be easy.

As the situation of the Battalion, as above mentioned, was extremely advantageous by having their flanks secured; let us now suppose one less so, by having one flank exposed, besides the front.

We will suppose then, that one of your flanks, as well as your front, lies open to the enemy; and that the horse have made a disposition to attack you in both. The only expedient against it is, the forming of two fronts, making the figure of an L, which is immediately done by wheeling back half of the Battalion, or a sufficient number of platoons, a quarter of the circle.

If you are to maintain that post, this disposition is better than an entire square, by having double the fire in each of these faces to those in the square.

The firing by platoons may be preserved in this figure as well as in Battalion, by dividing the platoons of each face into two firings, and reserving the whole front-rank for the third and last. If you think that the angle, where the two faces join, is exposed, a small platoon of Grenadiers may be formed on it: and that the platoons may have nothing to obstruct their firing, it will be proper to send the ensigns with the colours into the rear.

The fire of each face must be managed according as they are attacked; and no more platoons must be fired than what are absolutely necessary to repulse them, preserving the rest with the utmost care.

I own that I never heard of a single Battalion being formed into this figure upon action; and therefore

fore I shall not insist much on it; but as the flank of an army is often secured in this manner, by wheeling back of battalions and squadrons, I thought it might fall out the same way with a Battalion.

ACTICLE II.

I shall now proceed to shew how the square is to be formed from the said plans, without altering the former disposition of Officers, or any new telling off; and that it may be comprehended with the more ease, I have marked the platoons, which form each face, different from one another. See the annexed plans.

The figures in the front, are only the platoons numbered, from one to sixteen; by which you will see how they fall into the several faces of the square.

The figures in the rear, are to shew what firing the platoons belong to when formed in the square.

The platoons numbered (1.) on the inside, being on the right of each face, when faced square, which is outward, belong to the first firing.

Those numbered (2.) on the inside, belong to the second firing.

Those numbered (3.) on the inside, belong to the third firing.

Those numbered (4.) belong to the fourth firing.

The platoons of Grenadiers are likewise sub-divided for their forming on the several angles, their numbers shewing the angles on which they are to form.

I believe I need not give a further explanation of the plan, than what is already mentioned, for its being fully comprehended; so that I may proceed to give the proper directions for the forming the Battalion into the square, and reducing the square into Battalion.

As the Officers are not to be changed (but to remain in the posts assigned them, both in the front and rear, for the firing in Battalion) or any new Division

of

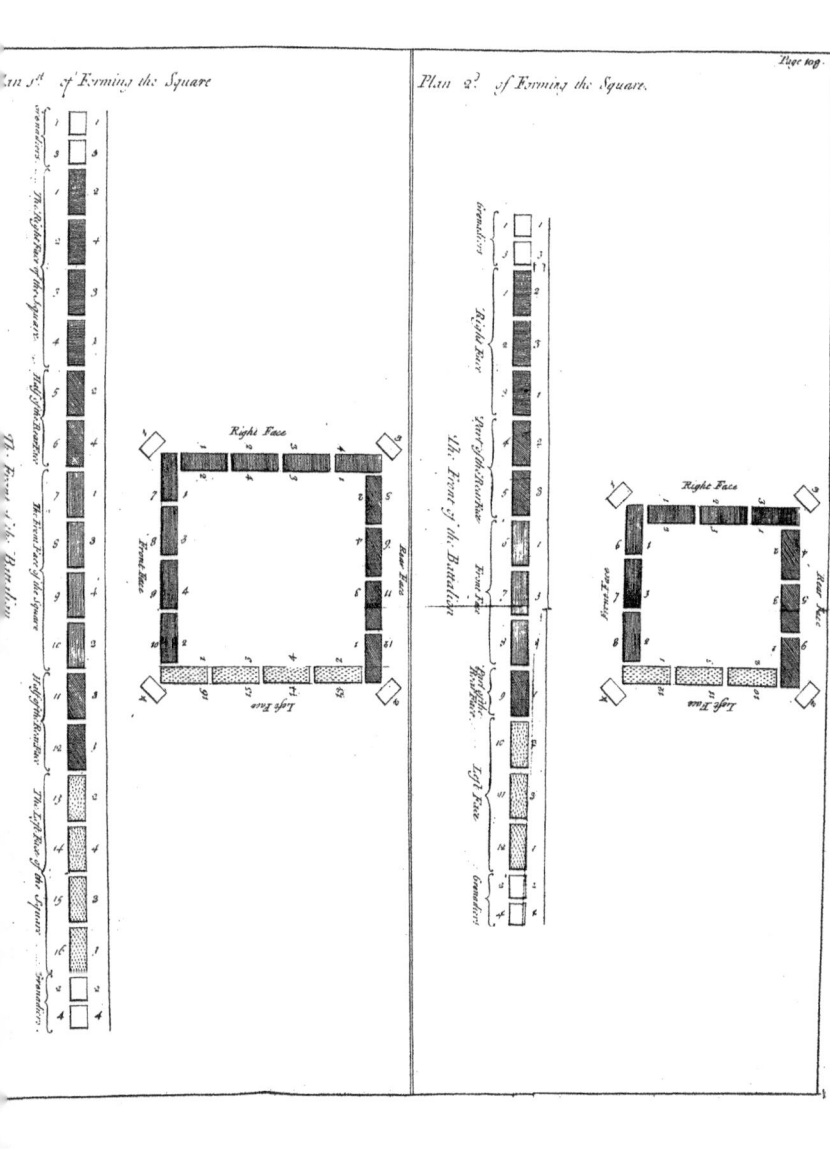

of the platoons, the square may be formed in a very short space of time; for which reason the commanding officer may defer the doing of it, till he sees the disposition actually made to attack the Battalion in every part.

As soon as he perceives this, he must avoid the danger, by forming the Battalion into a hollow square; or, according to the *French* way of calling it, *un Battalion quarré*, a square Battalion.

It is performed by three words of command, or by beat of Drum, as follows.

Take care to form the square by files! or a *Ruffle*.

    I. *Form the square*, or *a Flam*.
    II. *March!* or *a long Roll*.
    III. *Face square!* or *a Flam*.

In the following pages I shall explain what they are to perform at each.

At the first word of command, *form the square!* or *Flam*, the two platoons of Grenadiers, numbered 1 and 3; and the six platoons of the Battalion on the right, numbered 1, 2, 3, 4, 5, and 6, face, at two motions, to the left, the Officers and Serjeants, in the front and rear of those platoons, doing the same.

The six platoons of the Battalion on the left, numbered 11, 12, 13, 14, 15, and 16; together with the two platoons of Grenadiers, numbered 2, and 4; as also all the Officers and Serjeants belonging to them, face to the right.

The four platoons in the center, numbered 7, 8, 9, and 10, stand fast.

At the second word of command, *March!* or the *long Roll*, the six platoons on the right of the Battalion, file off to the left, and march in a straight line to the rear, till the leading platoon, numbered 6, has gained so much ground, as will be sufficient for the four

four platoons, that are to compose the right face, to form upon; upon which it faces to the right, by files, and keeps moving on, followed by the succeeding platoon, numbered 5, so far to the right, as just to leave room for it to draw up in its rear, thus together forming the left-half of the rear face, in the manner represented in the plan.

The four platoons in the center, stand fast, and form the front face.

The six platoons on the left of the Battalion, file off to the right, and march in a direct line to the rear, as did those from the right of the Battalion, the two leading platoons facing to the left by files, when they come to their proper ground, and then moving on till they form the right-half of the rear face.

The two platoons of Grenadiers, numbered 1 and 4, march to the right and left flanks of the four center platoons, which stood fast, and there wait, faced to the right and left inwards.

The two Grenadier platoons, numbered 3 and 2, step off with those of the Battalion, but march in a direct line, towards their respective angles, as marked in the plan.

The numbers here referred to, are those marked in the front of the Battalion.

After the foregoing word of command is executed, the Regiment will appear in this position.

The four center platoons, which compose the front face of the square, face outward to their proper front.

The eight platoons, which compose the right and left faces, face to the rear.

The two platoons which form the left-half of the rear face, face to the right, and those which form the right-half of the rear face, face to the left.

The two platoons of Grenadiers, numbered 2, and 3, which are to form on the rear angles, look into the rear face.

The

Chap. VII.   *Military Discipline.*   111

The two platoons of Grenadiers, number 1, and 4, which are to form on the front angles, face the flanks of the front face, looking along the ranks.

The Officers and Serjeants face as the platoons do, on which they are posted.

At the third word of command, *Face square!* or *a Flam,* they all face outward; the right face, and two left platoons of the rear face, facing to the left: the left face, and two right platoons of the rear face, facing to the right.

The two platoons of Grenadiers, 11, and 4, face to the front, the other two, numbered 2, and 3, to the rear; immediately after which, the whole four wheel up, and form on their respective angles, as described in the plan.

The Colonel, Major, Ensigns with the Colours, Adjutant, and Drummers, march into the square; the Drummers drawing up in the center, in the rear of the Colours.

The Field-Officers can have no fixed post assigned them in the square standing; but are to have a watchful eye over the whole, and to move about from place to place, to give the necessary directions to the several parts, as occasion may require.

Whenever the square is to march, the Ensigns with the Colours, and the Drums, are immediately to repair to that face, by which it is to march; and the commanding Officer is to come out of the square, and lead it: the Lieutenant-Colonel likewise, is to be on the out-side, to bring up the rear.

The square being on the march, as soon as the Grenadiers have fired, they are to come up to their *Recover,* face to the left, and march nimbly into the square, the right-hand platoon of each face, having recovered their arms, and made a half-wheel to the left, to let them in. They are always to form again, behind their right-hand platoons, and then proceed to load; being never to load on the out-side of the square.   The

The Officers who command the platoons remain in the front rank with them, and those who were posted in the rear remain within the square, in the rear of the several platoons; and when any of the Officers in the front are killed or wounded, the Officers in the rear of those platoons are to move out immediately, and take the command.

It is evident from hence, that the square may be formed in this manner, in less than a minute, if pressed in time; there being no alteration required in the disposition of Officers from that of the Battalion drawn up for action; or any new telling off the platoons. Besides, another advantage in this manner of forming the square, is, that you preserve a front of the four center platoons without moving, which will secure you till the square is formed; a circumstance, in my opinion, of no small consideration.

As victory, even in a superior army, is uncertain, from the variety of circumstances incident in action: and when we imagine fortune hovering over us with a crown of laurel, she often eludes our hopes, and bestows it on the adverse party; we must not therefore depend on her too much, but act with caution, and be prepared against all events, before we enter upon action. And as the making a handsome retreat is the most difficult part of the service, and, next to the gaining of a battle, the most commendable; it is therefore incumbent on the commanding Officer of every Regiment to have the same regard towards the preservation of his men, as the General has to the whole. For which reason, the platoons of every Battalion should be told off, in such a manner, and the Officers appointed to them, that when the Battalion is ordered, or forced to retire, it may be performed without any further directions than the words of command, for the marching off in Battalion, by grand or sub divisions, by platoons, or in the square; by which different ways the Battalions should be told off,

Chap. VII.   *Military Discipline.*   113

off, and the Officers and Soldiers thoroughly acquainted with them before they engage, that, when ordered, they may be in no confusion in the performance.

How the square is to be reduced.

The square may be reduced into Battalion with as much ease, and in as short a space of time, as it was formed in; for the performing of which there are only the three following words of command, or signals from the drum.

Take care to reduce the square! or *a Ruffle.*

I. *From the square form the Battalion!* or *a Flam.*

II. *March!* or *the Long-roll.*

III. *Halt!* or *a Flam.*

At the first word of command, or *Flam*, the four platoons of Grenadiers wheel backwards, dress in a line, with the front and rear faces, and face outwards.

The six platoons, which form the right face, and the left of the rear face, face to the left.

The six platoons, which form the left face, and the right of the rear face, face to the right.

At the second word of command, *March!* or the *Long-roll*, the whole step off by files, and move briskly, to their former ground.

The Colonel, Lieutenant Colonel, Major, Ensigns with the Colours, Adjutant, and Drummers, march at the same time, to their respective posts.

As soon as all the platoons are come up in a line with the front-face, the Colonel, or commanding Officer, is to proceed to the third word of command, *Halt!* or *a Flam.*

At this, the whole face at two motions to their proper front; by which the square is reduced, and the Battalion formed as before, without moving the Officers from their platoons, either in the front or rear, in the forming or reducing the square.

The first plan being calculated for a Regiment consisting of six hundred men, left it should be reduced to about five hundred, I have annexed a second plan accordingly, told off into twelve platoons besides Grenadiers; the forming of which into a square, and the reducing it again into Battalion, is to be performed in the same manner as the first; with this difference only, that as each face is composed but of three platoons, there must be two platoons taken from one flank, and one platoon from the other, to form the rear face; which difference is so small, that I believe it will not be objected against in service, whatever it may in exercise.

When the strength of the Battalion will allow of it, the telling off the platoons, according to the first plan, is what I would recommend, as being the most perfect both for the firings in Battalion and in the square, and for the regularity and ease in forming the square. If the Regiment consists of six hundred men, they may be told off into eighteen platoons, including the two platoons of Grenadiers, two platoons of which will consist of twelve files each; and the other sixteen platoons of eleven files each; so that by making of the two platoons of Grenadiers twelve files each, they will have, when they are subdivided for the forming of the square, a platoon of six files for each angle, which is as few as they ought to have for the covering them.

But when a Battalion consists of five hundred men, the model of the second plan may be followed, most of the platoons of which will consist of twelve files; they may be divided into three firings, containing four platoons in each, and the Grenadiers kept for the reserve, which should be carefully preserved if your

flanks

Chap. VII. *Military Discipline.* 115

flanks are the least exposed to the enemy's attacks; and if you are under any apprehension of the enemy's horse, it would be very proper to strengthen your reserve, by adding the whole front rank of the Battalion to it, and only fire the two rear ranks of the three firings by platoons.

ARTICLE III.

Manner of forming the square by four grand divisions, according to the plan annexed.

When the square is to be formed by four grand divisions, without having gone through any part of the platoon firing, they are to proceed in the following manner.

The rear half files of the Battalion must be doubled to the left; after that the Grenadiers must be divided on the right and left, and then sub-divided for the angles.

The Battalion must be divided into four grand divisions, and each grand division sub-divided into three platoons each. See the annexed plan, where the said is told off, as here described.

The Captains, Subalterns, and Serjeants, are to be divided equally on the four grand divisions; after which there must be an Officer appointed to command each platoon, who continue in the front; but the remaining part of the Officers are to fall immediately into the rear of their several grand divisions.

When the divisions are told off, and the Officers appointed to them, as above directed, the commanding Officer is to proceed to the words of command, or signals from the Drum, for the forming of the square, which are the same as in the other way of doing it.

I. *Form*

I. *Form the square!* or *a Flam.*

At this word of command, the firſt grand diviſion and the two platoons of Grenadiers, number 1, and 4, face to the right and left inwards, thus; the platoon of Grenadiers, number 1, and the firſt grand diviſion face to the left, and the platoon of Grenadiers, number 4, faces to the right.

The other three grand diviſions, and the two platoons of Grenadiers, number 2, and 3, face, at the ſame time, to the right-about.

The Officers, Serjeants, and Drummers, face as the ſeverel diviſions do, on which they are poſted.

After this the commanding Officer proceeds.

II. *March!* or *the Long-roll.*

After this word of command, the whole are to march and form the ſquare, thus:

The ſecond and fourth grand diviſions wheel inward a quarter of the circle, and form the right and left faces of the ſquare.

The third grand diviſion, with the Colonel, Lieutenant-Colonel, Enſigns with the Colours, and the third diviſion of Drummers, march in a ſtraight line to the rear, till they come to the extreme flanks of the right and left faces, and then ſtand, which third grand diviſion forms the rear face of the ſquare.

The firſt grand diviſion marches to the left, till they come into the ground where the third grand diviſion ſtood, and then ſtand, being to form the front face of the ſquare.

The platoon of Grenadiers, number 2, wheels with the left face, and ſtands when they do, being to cover that angle.

The platoon of Grenadiers, number 3, marches to the right flank of the right face, and ſtands, being to cover that angle.

The

Chap. VII.   *Military Discipline.*   117

The two platoons of Grenadiers, number 1, and 4, being faced inwards, march in a direct line to the flanks of the front face, and then stand, being to cover the front angles.

When the several grand divisions and platoons of Grenadiers have marched as above directed, they will appear in the following position.

The first grand division, composing the front face of the square, stand faced to the left.

The third grand division, forming the rear face, face to the rear.

The second and fourth grand divisions, which form the right and left faces of the square, face into the square.

The two platoons of Grenadiers, number 1, and 4, being to cover the front angles, face to the right and left inwards.

The two platoons of Grenadiers, number 2, and 3, being to cover the rear angles, face as the right and left faces do.

The Officers and Serjeants face as their respective divisions do.

The Colonel, Lieutenant-Colonel, Ensigns with the Colours, and Drummers, having marched as before directed, fall into the square.

As soon as they have come to their ground, and stand, the commanding Officer proceeds to the third and last word of command.

III. *Face square!* or *a Flam.*

At this word of command, the whole face outward, thus:

The front face going to the right, and the right and left faces to the right-about, the two platoons of Grenadiers, number 1, and 4, face to their proper front, and the two platoons, number 2, and 3, face to the right-about; immediately after which, the four platoons of Grenadiers wheel back and cover their several angles.

The Officers and Serjeants face as their divisions do, and the Drum-Major is to draw up the Drummers, in the rear of the Colours, in the center of the square.

The Officers, who were posted in the rear of the third grand division, are to move immediately into the square; and the Officers, who were appointed to command the platoons of that division, are to move out, the rear rank of that face becoming then the front.

As soon as they have faced square, the Major and Adjutant march into the square, no officer remaining without, but those who command the platoons.

The firings in this are the same as that of the second plan, they being calculated for the same number of men; but if the Battalion consists of six hundred men, the grand divisions may be divided into four platoons each, as the first plan is, and yet keep to the forming the square by grand division.

I shall now shew how it is to be reduced into Battalion.

I. *Reduce the square!* or *a Flam.*

At this word of command, the whole being faced square, the front face, or first grand division, faces to the right; the rear face, or third grand division, faces to the right-about; the two platoons of Grenadiers, number 1, and 4, covering the front angles, wheel towards their proper front, and when they come in a line with the front-face, they are to face to the right and left outward; the two platoons of Grenadiers, number 2, and 3, covering the rear angles, wheel towards their proper front, till they come in a line with the right and left faces, or second and fourth grand divisions, and then stand. The Officers face with their divisions.

II. *March!*

II. *March!* or *the Long-roll.*

At this, the whole march and form the Battalion, thus:

The front face and the first platoon of Grenadiers march in a direct line to the right, and as soon as they have got to the right of the right face, they are to stand; only the first platoon of Grenadiers is to march a little further, that the third platoon of Grenadiers may have room to form between them and the right of the first grand division.

The right and left faces wheel towards their proper front a quarter of the circle, and then stand, the second platoon of Grenadiers wheeling up on the flank of the left face, or fourth grand division.

The third platoon of Grenadiers marches to the right, and forms between the first platoon of Grediers, and the right of the Battalion.

The fourth platoon of Grenadiers marches in a straight line to the left; and when they have left room enough for the left face and second platoon of Grenadiers to form in, they are to stand.

The rear face, or third grand division, with the Colonel, Lieutenant-Colonel, and Ensigns with the Colours, march straight forward to the front, and when they come between the second and fourth grand divisions, and dress in a line with them, they are to stand.

As soon as they have all got into their proper posts, as before, the commanding Officer proceeds.

III. *Halt!* or *a Flam.*

At this word of command, they all face to their proper front, thus: The first platoon of Grenadiers and the first grand division face to the left, and the fourth platoon of Grenadiers to the right; after which the Officers in the rear may be ordered into the front, and the Drummers to their former posts; which completes the reduction of the square into Battalion.

The only thing that is irregular in the forming of the square in this manner, is in the third grand division, by the rear rank becoming the front, and the front rank the rear, when the square is formed, and the Officers in the front and rear changing of their posts; but this piece of irregularity is of no great consequence upon service, since the men in the rear rank may be as good as those in the front, and the Officers may change in a moment.

But the greatest fault consists in there being no front preserved while the square is forming, the whole being in motion at the same time, which may be of dangerous consequence, if the enemy's horse should be near.

Whereas the other manner of forming the square, as explained in the second article, has not the irregularity above-mentioned, nor the danger, while it is forming, for want of a front to the enemy. Besides, it may be done quicker, and with as much ease, by practising of it at exercise, as that by grand divisions: however, those who do not approve of that way, (which I imagine will be but very few) may follow the other; but before they determine absolutely, it will be but fair to try both.

I shall give directions in the following article, how they are to fire and march in the square at exercise, since the doing of it upon action must depend on the manner you are attacked, in which the commanding Officer must be directed by his own judgment and experience.

ARTICLE

ARTICLE IV.

Directions for marching and firing in the square.

The square being formed, and the platoons of each face divided into their proper firings, as described by the different plans in the foregoing article, they are to proceed to the firings.

I shall begin with directions for that of the first plan, containing four firings, besides the Grenadiers, which is one more than they are generally told off in, in Battalion, which renders it, in my opinion, the more perfect, as being of greater service than when they are divided into three.

For when they are told off into three firings, whether in Battalion, or in the square, the entire front rank is commonly kept for the reserve, and the two rear ranks only fired by platoons; so that in reality there are four firings, without being called so: though I humbly conceive, the effect will not be the same; from the observation I made on the firing by ranks in the fifth article of the preceding chapter; as also in the first article of this, on reserving the front rank.

Those who differ with me on this head may divide the square into three firings, according to the method of the second plan; but as the four firings will appear better in the exercise, I will pursue that scheme, and give the necessary directions accordingly.

The firings may be performed standing, or by making a movement before each firing.

In firing the square, the commanding Officer is always to apprise the Battalion, whether it is to be by platoons, or by faces, before he makes the signal to begin.

The four platoons of Grenadiers on the angles, are to fire first; after which they are immediately to retire into the square, and form in the rear of the right-hand platoons, in order to load, as is above directed.

The

The firſt firing of the ſquare, conſiſts of the right hand platoon, of each face, numbered 1, on the inſide.

The ſecond firing conſiſts of the left-hand platoon of each face, numbered 2, on the inſide.

The third conſiſts of the right-hand center platoon of each face, numbered 3, on the inſide.

The fourth conſiſts of the left-hand center platoon, numbered 4, on the inſide.

When you would go through the firings both ſtanding and marching, it will be proper to vary them from one another.

Thoſe ſtanding, to be performed in their order; and thoſe marching, together; as is explained in the third article of the ſixth chapter.

But leſt firing in their order in the ſquare ſhould not be thoroughly comprehended by the directions in the article above-mentioned, we ſhall here explain how it is to be performed.

When the Grenadiers are to fire in their order, the firſt platoon of Grenadiers, number 1, covering the angle on the right of the front face, fires firſt. The platoon of Grenadiers, number 2, on the right of the rear face, fires next. The Grenadiers, number 3, on the right of the right face, fire the third. The Grenadiers, number 4, on the right of the left face, fire laſt.

When the platoons of the ſquare are to fire in their order, if it is thoſe of the firſt firing, they are to make ready together, and the platoon, number 1, of the front face, fires firſt; then the platoon, number 1, of the rear face, fires next; after that, the platoon, number 1, of the right face; and, laſtly, the platoon, number 1, of the left face.

The platoons of the other firings are to obſerve the ſame method, when they are to fire in their order, by beginning with that in the front face; ſecondly, that of the rear face; thirdly, the right face, and, fourthly, the left face,

The

The firſt firing, after the Grenadiers, conſiſts of the four platoons of the ſquare, number 1, on the inſide.

The other three firings conſiſt of four platoons each, one in each face of the ſquare. See how they are diſpoſed of by the plan, the numbers on the inſide ſhewing which firing they belong to.

How the ſquare is to fire ſtanding.

At the beating of the *Preparative*, the four platoons of Grenadiers make ready together, and the Officer, who commands the firſt platoon of Grenadiers, number 1, gives the words to *preſent* and *fire*; after which they are to recover their arms, face to the left, and march into the ſquare by files, led by the Serjeant, the Officer bringing up the rear; there halt in the rear of the right-hand platoon, and proceed to load.

As ſoon as the firſt platoon of Grenadiers has fired, the Officer commanding the ſecond platoon of Grenadiers is to order his to do the ſame, and then march into the ſquare. After that the third platoon of Grenadiers is to fire, and then the fourth.

When a platoon of Grenadiers has the word to *preſent*, the right hand platoon upon that angle, is to recover their arms, and wheel out, in order to open a paſſage for them to retire into the ſquare, as ſoon as they have fired, immediately after which it is to fall back again, and ſhoulder. When the four platoons of Grenadiers have fired, a ſecond *Preparative* is beat; upon which the right-hand platoons, which compoſe the firſt firing of the ſquare, are to make ready, and fire in their order.

After the firſt firing is over, a *Preparative* is to be beat for the platoons of the ſecond firing to make ready, and fire, which they are to do in their order.

The

The third and fourth firings are to be performed in the same manner.

As soon as the platoons have fired, they are imdiately to load and shoulder.

After the last firing is over, the exercising Officer orders the Drummers to beat the *General*, as a signal for the firing to cease.

After that, a *Ruffle*, for the four platoons of Grenadiers, to prepare to cover the angles of the square again: upon which they, as also the right hand platoons, are to recover their arms. Next to that a *Flam*, at which the right-hand platoons wheel to the left, and the Grenadiers are led out by their Officers, forming instantly upon the angles, as before, and the right-hand platoons falling back to their places.

The next *Preparative* is for the four platoons of Grenadiers to make ready again. The exercising officer gives them the word of command to *present* and *fire* together, after which they recover their arms, and retire into the square as before, the right hand platoons wheeling to the left, to let them in.

The Grenadiers having fired together, and formed again in the rear of the right-hand platoons, a *Preparative* is beat for the whole square to make ready, and to be fired by faces, by word of command from the exercising officer, beginning with the front face; then the rear face; after that the right face; and, lastly, the left face; each face loading, and shouldering their arms, as soon as they have given their fire.

When the left face has fired, the *General* is beat, to signify that the same fire is not to be repeated, as has been before observed.

The whole square being shouldered again, a *Ruffle* is to be beat, for the Grenadiers to march out again, they, and the right-hand platoons, recovering their arms. Then a *Flam*, at which the platoons wheel, and the Grenadiers move out, as before.

Chap. VII. *Military Discipline.*

At the next *Preparative*, the Grenadiers make ready again, and are fired together, by word of command from the exercising Officer ; after which they retire, as usual, into the square.

Then follows a *Preparative* for the whole square to make ready, and to be fired together, by the exercising Officer.

The fire being terminated by beating the *General*, and the square shouldered again, a *Ruffle* is beat, at which the Grenadiers, and right-hand platoons, recover their arms. At the *Flam*, the Grenadiers march out, and cover their respective angles, as before directed.

The Battalion having gone through the firings in the square standing, I shall now shew how they are to be performed marching.

Directions for the square to march.

Whenever the square is to march, the commanding Officer is first to apprise the Battalion, whether it is to march by the front face, by the right face, by the left face, or by the rear face: upon which the Colours and Drums are to repair to that face, as has been before observed, and the commanding Officer is to come out of the square, and lead it : the Lieutenant-Colonel likewise is to be on the outside, to bring up the rear.

As there are four firings told off in the plan of this square, they should make a movement towards each front before each firing, by marching twenty or thirty paces at a time, or more or less, as the commanding Officer shall think proper, or the ground admit of it.

Before we proceed further, it will be necessary to give directions how the several parts of the square are to face and march, on the Drummers beating on the different fronts.

When the Drummers are ordered to beat a *March*, (which is to be the Grenadiers *march*) in the rear of
the

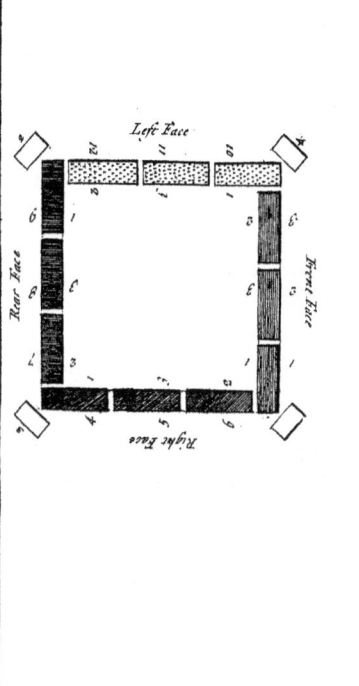

Plan 3d of Forming the Square by Grand Divisions standing.

the front face, the whole square is to face and march towards that front, thus: The rear face comes to the right-about; the right face goes to the left; the left face to the right, and the four platoons of Grenadiers wheel an eighth part of the circle towards that front. All the Officers and Serjeants, are to face as those parts do on which they are posted.

As soon as they have faced, as above directed, they are all to step forward together, and march in that order, without opening or closing their ranks or files, very slow towards the said front as long as the *March* is beat, and when the Drummers are ordered to cease, they stand fast, waiting for a *Flam*, at which they face square, thus:

The rear face goes to the right-about.
The right face to the right.
The left face to the left; and,
The four platoons of Grenadiers wheel back and cover their angles.

When the Drummers beat the *Grenadiers March* in the rear of the rear face, they are all to face and march toward that front, thus:

The platoons of Grenadiers wheel, as before, towards that front.
The front face goes to the right-about.
The right face faces to the right; and,
The left face faces to the left.

After which, they are all to march towards the rear front, as long as the *March* continues beating; and when it ceases, they are to stand fast, till the *Flam* is given, at which they are to face square thus:

The Grenadiers are to wheel back and cover their angles.

The

Chap. VII. *Military Dicipline.*

The front face goes to the right-about.
The right face to the left; and,
The left face to the right.

On the beating of the *Grenadiers March* in the rear of the right face, they are all to face and march towards that front.

The Grenadiers are to wheel towards that front.
The left face goes to the right-about.
The front face to the right; and,
The rear face to the left.

As soon as the *March* ceases, they are all to stand fast; and, at the *Flam*,

The Grenadiers wheel back and cover their angles.
The left face goes to the right-about.
The front face to the left; and,
The rear face to the right.

On the beating of the *Grenadiers March* in the rear of the left face, they are all to face and march towards that front.

The Grenadiers wheel towards that front.
The right face goes to the right-about,
The front face to the left; and,
The rear face to the right.

When the *March* ceases, they are all to stand fast; and, at the *Flam*,

The Grenadiers wheel back and cover their angles.
The right face goes to the right-about.
The front face to the right; and,
The rear face to the left.

The

The several faces of the square are to face to that front, towards which they are to march, at the first stroke of the *Grenadiers March*: and it is to be beat by all the Drummers, drawn up immediately in the rear of the Colours; which finishes the directions how the several parts of the square are to face and march towards each of the four fronts.

The square being to march before each firing, in order to vary it from that standing, the platoons of each firing should fire together; if so, the commanding Officer must acquaint them with it, and give the words of command himself: after which he proceeds in the following manner:

Grenadiers take care to fire a volley!

He then orders a *Preparative*, for them to make ready, and gives the words,

Present! Fire!

As soon as the Grenadiers have retired into the square after firing, and are loaded, and shouldered again, he gives the caution for the square to march by the front face, and then orders the Drummers to beat the *Grenadiers March*; upon the first stroke of which, the whole square faces, as above directed, and marches very slow towards the front of the front face: and when it has marched as far as shall be thought proper, he is to order the Drummers to cease beating, at which the whole stands fast, and waits for a *Flam* to face square.

After this, he orders a *Preparative*, on which the four platoons of the square, of the first firing, make ready. That done, he gives the words,

Present! Fire!

After

After the first firing is over, the commanding Officer proceeds thus:

Take care to march by the rear face!

At this the Colours and Drums repair to the rear of the rear face; and upon their beginning the *Grenadiers March*, the whole square faces, and marches, as above directed, towards the front of that face. When the Drums cease, they halt; and, at the *Flam*, face again to their proper front.

He then orders a *Preparative* for the platoons of the second firing, which having made ready, he proceeds:

Present! Fire!

The second firing being over, he goes on:

Take care to march by the right face!

The Colours and Drums repair to the rear of the right face: upon beating the *Grenadiers March*, the square faces, and marches towards the front of that face. When the Drums cease, and the *Flam* is beat, the whole face to their proper front, and a *Preparative* is ordered for the platoons of the third firing to make ready. The exercising Officer then gives the word, as before,

Present! Fire!

When the third firing is over, the caution follows:

Take care to march by the left face!

The Colours and Drums having placed themselves in the rear of that face, and the *Grenadiers March* beating, the square faces, and marches that way. After the *Flam* to face square, the *Preparative* is beat for the fourth firing, which being made ready, receives the word of command,

Present! Fire!

As soon as the fourth firing is over, the Grenadiers are to march out of the square and form on the angles, as before directed, in that part, where they fire standing.

When the platoons of each firing have fired, they are to load in marching, and then shoulder.

This completes the several movements of the square, with a firing after each; which firings may be varied, if the commanding Officer thinks proper, from the foregoing, after this manner:

After the first movement, the whole front face may be ordered to fire together.

After the second movement, the rear face fire together.

After the third movement, the right face fire together.

After the fourth movement, the left face fire together.

This manner of firing may appear very well in the exercise, but would prove too dangerous in service: since there would be an entire front for some time, without any fire to defend themselves: therefore I only mention it as proper for the exercise, to vary it from the other.

They may likewise fire by ranks, as they do in battalion, by ordering the whole to make ready together, then begin with the rear rank of the whole square; after that the center rank of the whole; and, lastly, the front rank.

After

After they have gone through the different firings of the square, both standing and marching, they may finish with a fire of the whole square, which may be done where they stand, or after a movement towards their proper front.

As soon as this fire is over, the square should be reduced into battalion; the rear ranks opened, the Officers in the rear ordered to the front, and to take their posts in battalion; the Grenadiers on the left ordered to the right; the bayonets unfixed, and the rear half-files to the right as they were; which does not only conclude this chapter, but also the directions for the different firings of the foot in every part of the service: but as what I have hitherto mentioned is only the rudiments of discipline, I shall endeavour to raise the subject, by treating on such parts of the service, as will give all those who are desirous to know it, a general notion of their duty, on different occasions.

CHAP. VIII.

Consisting of general rules for the marching of a Regiment of foot, or a Detachment of men, where there is a possibility of their being attacked by the enemy.

ARTICLE I.

THERE is not any thing in which an Officer shews his want of conduct so much, as in suffering himself to be surprised, either upon his post, or in marching with a body of men under his command, without being prepared to make a proper defence, and by not having taken the necessary precautions to prevent it.

When an Officer has had the misfortune of being beat, his honour will not suffer by it, provided he has done his duty, and acted like a Soldier. But if he is surprised by neglecting the common methods used to prevent it, his character is hardly retrievable, unless it proceeds from his want of experience; and even in that case he will find it very difficult.

An Officer, who is detached with a body of men, ought to consider, that the lives of those under his command depend in a great measure on his prudence; and if he has any important post committed to his charge, the lives of many more may follow.

This consideration alone, without mentioning the loss of reputation, is sufficient, in my opinion, to make us apply ourselves to our duty with a more than
com-

common zeal, that we may not be ignorant in what relates to our profession, when our King and Country has an occasion for our service.

The military profession has, in all ages, been esteemed the most honourable, from the danger that attends it. The motives that lead mankind to it, must proceed from a noble and generous inclination, since they sacrifice their ease, and their lives, in the defence of their country.

To answer this glorious end, we should endeavour at the knowledge of our calling, by a thorough application to the service.

The same spirit that brings us into the army, should make us apply ourselves to the study of the military art, the common forms of which may be easily attained by a moderate application, as well as capacity: neither is it below any military man, let his birth be ever so noble, to be knowing in the minute parts of the service. It will not cramp his genius (as some have been pleased to say, in order, as I suppose, to excuse their own ignorance) but rather aid and assist it in great and daring enterprizes.

Our great and warlike neighbours the *Germans*, are so entirely prepossessed in favour of this opinion, that they oblige even their youth of quality to perform the function of a private Soldier, Corporal, and Serjeant, that they may learn the duty of each, before they have a commission: and sure no nation has produced greater Generals.

Our late monarch, the glorious King *William*, whose military capacity was second to none, was perfectly knowing in the small, as well as the grand detail of an army. In visiting the out-posts, he would frequently condescend to place the sentinels himself, and instruct the Officers how to do it. He was a strict observer of all the parts of discipline: and knew the duty of every one in the army, from the highest to the lowest: and if so great a Prince thought it a necessary ceffary qualification, I believe there will be hardly any one found of another opinion.

I do not pretend to infer from the above obfervations, that it is abfolutely neceffary for our young nobility and gentry to pafs through thofe little and fervile Offices before they arrive at a commiffion; but I think it abfolutely neceffary that they fhould apply themfelves to the fervice, as foon as they have one: for without they know the duty of thofe under their command, how can they pretend to direct?

A commiffion, it is true, qualifies a man for the pay: but it muft be time and experience, and a thorough application to the fervice, that entitles him to the appellation of a Soldier.

He that makes himfelf mafter of the duty of thofe below him, will the eafier comprehend what is due to thofe above him; and be a means to qualify him for a higher poft, and to do the duty of it with honour and credit when given him; with this addition, that he was fit for the poft, and not, that the poft was fit for him.

It is more commendable and praife-worthy to owe our preferment to merit than favour. The dependance on the latter, is the reafon why fo many young gentlemen neglect the former.

Money and powerful relations will always procure them what they want; they have therefore no occafion to apply themfelves to the knowledge of their duty. It is from this way of thinking that fo many of them do fo little credit to their pofts; not from the want of genius, but application.

I hope thefe few obfervations will not be taken as a reflection on the young gentlemen who have come lately into the army; but rather as an admonition to avoid the neglect complained of; my defign being purely to ferve them, that they may be the better qualified to ferve their country when fhe calls upon them.

A R.

ARTICLE II.

I shall now proceed to what was proposed in this chapter, the necessary precautions proper to be taken in the marching of a Regiment, or a Detachment of men, to prevent a surprise, &c.

When a Regiment is to march through a country, where there is a possibility of meeting with the enemy, the commanding Officer should leave nothing to chance. Fortune may fail us, if we trust too much to her; but a prudent conduct never will. It is true, we may be overpowered and conquered, notwithstanding all our care; but never shamefully beat, if we act as we ought: and a man may gain reputation, though he is overcome.

The common method of marching a Regiment, is by platoons, as described in the sixth article of Chapter VI. on street-firing.

Marching in this order, the Regiment will be ready to enter upon action in whatever shape it may be required, whether in street-firing, in battalion, or in the square.

But there is another advantage, besides that above-mentioned, by the Officers being divided to, and marching with the platoons, which is, that the men will march in greater order by having the eyes of so many Officers on them, nor venture to leave their ranks without leave, for fear of being discovered; a consideration, I am sure, of no small consequence, since it will be the means to prevent a great many men from being killed by the country people, either in the defence of their goods, or out of hatred to the Soldiers; or from being taken by partisan parties, when they fall behind: but though they should have the good fortune to escape both, the apprehension of being punished, by quitting their posts contrary to orders, too often induces them to desert.

The common method used to prevent your being attacked on the march before you have time to make a proper defence, is, by having a van and rear-guard, which guards may be stronger or weaker according to the danger you may apprehend from the enemy, or the country you are to march through.

Those guards are generally commanded by Officers, and frequently by Captains. They should never lose sight of the Regiment, or at least be out of the hearing of the Drum; for which reason there should be a Drummer ordered to beat in the rear platoon or division, as well as in the front; but more particularly in night-marches.

Before the Regiment marches from the parade, or the head of their encampment, these guards are to be drawn out; and if your march is towards the enemy, or that you apprehend more danger in the front than the rear, your vanguard should consist of a Captain's command, and the rear-guard only of a Subaltern's: however, this depends on the discretion of the commanding Officer.

As soon as these guards are formed, the Officers who command them, should receive their instructions from the commanding Officer of the Regiment or Detachment. But as young Officers can have but a very imperfect notion of the intent and meaning of these guards, without some further insight than barely the mentioning of them; I shall endeavour to give them a clear idea of the nature and design of those guards, by setting down the duty of each.

General instructions to the van-guard.

The vanguard is to march before the Regiment. The distance which they are to be advanced, cannot be absolutely determined; since it must depend on the nature of the country you march through; so that in an inclosed country it can hardly exceed two hundred

Chap. VIII. *Military Discipline.* 137

dred yards without losing sight of the Regiment, which they are by no means to do, unless they have orders: and in an open or champaign one, they ought not to be above three or four hundred yards, left they should be attacked and cut off by a superior party, before the Regiment could come up to their relief.

The van-guard is to reconnoitre, or view, every place where any number of men can lie concealed, such as woods, copses, ditches, hollow ways, straggling houses, or villages, through which you are to march, or pass near.

That the Regiment may not halt upon every occasion of this nature, the Officer who commands the van-guard must order a Serjeant, and six or twelve men, to advance before him, but not to march out of his sight, who are to reconnoitre all suspected places; and where there are more than one of those places to be looked into at a time, by having them both on the right and left of the road, he is to order out another small party for that purpose.

When there are any woods or villages which will require some time to view, the Officer must halt his guard at some distance from them, and remain there, till his advanced parties have reconnoitred them thoroughly, and sent him an account that all is safe; after which he is to march on.

Upon every halt of this kind, he is to send one to the commanding Officer of the Regiment, to acquaint him with the reason of his halting; upon which he should halt the Regiment as soon as they come in sight of the van-guard; and when it marches again, the Regiment is to do so to.

The reason for the van-guard's halting at some distance from a wood or village till it is reconnoitred, is for fear of an ambuscade; for should they march up too near before it is viewed, they might be drawn too far into the snare to be able to extricate them-

selves,

selves, and, by that means, draw the Regiment into the same misfortune; whereas, by halting at some distance, that danger is avoided; at least so far, that they cannot surprise you, by falling upon you unprepared, which is all that can be expected from an Officer.

The same reason holds good for the Regiment's halting, when the van-guard does.

When the van-guard discovers any body of men, it is to halt, and the Officer is to send back immediately and acquaint the commanding Officer with it, and to know what particular commands he has for him: and when he discovers any thing further, he is to do the same, whether it relates to their numbers, quality, (as horse or foot) movement and disposition, that he may take his measures accordingly. Thus I have given as full an account of the duty of a vanguard as the nature of the thing will admit of, or that general rules can direct.

General instructions to a rear-guard.

The chief employment of a rear-guard is to take up all the Soldiers who shall fall behind the Regiment, and march them prisoners, in order to their being punished, for leaving it; which but too many will do, without a great deal of care, in order to plunder or marode.

This precaution is therefore absolutely necessary, without which a great many men may be lost, and the country suffer extremely, by being left to the discretion of those gentlemen.

The Officer commanding the rear-guard, must therefore be very diligent, in examining every place in which the Soldiers can hide themselves, to prevent these disorders.

As the rear-guard is not to be at any great distance from the Regiment, it will likewise prove a security, in preventing their being fallen upon in the rear, before they have notice to prepare for their defence: for the moment that any troops appear in the rear, the Officer of that guard muft fend and acquaint the commanding Officer with it, that he may have time to make a difpofition fuitable to the occafion; to gain which, the Officer of the rear-guard is to oppofe them in the beft manner he can; but if the fuperiority of the enemy obliges him to give way before he can receive further orders from the commanding Officer, he muft endeavour to join the Regiment by a flow and regular retreat, in making a ftand at every fpot of ground that can be difputed. If he finds it impracticable for him to join the Regiment, by his retreat being cut off, he muft endeavour to gain the neareft place of fecurity, whether inclofures, woods, hollow-ways, moraffes, villages, or towns, in order to fave himfelf and party; but this fhould not be attempted while there are any hopes left of his putting a ftop to the enemy, or his joining the Regiment, fince it will be weakened by his going off.

ARTICLE III.

Having explained, in the foregoing article, the nature and defign of the van and rear-guards, I fhall now proceed to what relates to the body of the Regiment, or Detachment, and in what manner the commanding Officer is to conduct himfelf for the fecurity of the whole.

A good deal of care and judgment is required in the marching of a Regiment in good order, and to prevent its running out into too great a length.

The Officers on the feveral platoons, or divifions, muft endeavour to avoid it, by keeping up their divifions to a proper diftance from thofe before them; for if they once fuffer the divifion in their front to march any confiderable diftance from them, they will not only fatigue their men, but find it a very difficult matter

matter to regain the ground so lost; and if it proves so to one division, it will be much more difficult to those in the rear: for which reason, the Officers cannot be too exact in the marching of their divisions, and the keeping of their ranks to their due distance, particularly in inclosed countries, where the roads are generally narrow; but if the Officers neglect this precaution, the Regiment may run into such a length, that the front may be attacked and beat, before the rear can be brought up to sustain them. But lest the Officers should fail in this part of their duty, the commanding Officer should order the Major and Adjutant to halt by turns every half hour, to see the Regiment pass by, and bring him an account in what order they find it, that he may direct his march, by moving faster or slower, according to the report made him.

When the Regiment is to pass a defile, where a division cannot march entire, but are obliged to rank off, that is, to pass by half or quarter ranks, the Officer, who leads the first division, should halt, or march extremely slow, after he has passed it, till he has an account that they are all over, and come up to their proper distance, upon which notice he is to march on as before.

The Officers must make the men of their respective divisions pass the defile as fast as they can without loss of time, and fall into their ranks the moment they are over; and when their divisions are formed, they must march them as quick as possible, without running, till they join those in their front, and then march as they do.

If the above directions are observed, the Regiment will not only march more secure, but also quicker, and with less fatigue to the men; but if they are neglected, it will be impossible for the rear divisions to keep up, so that in half a day's march the Regiment

may

Chap. VIII.   *Military Discipline.*   141

may extend itself, from front to rear, a mile or two, which may prove of dangerous consequence.

Particular care should be taken that the Soldiers do not fasten the tent-poles to their firelocks, (which is frequently done for the ease of carrying them) left they should be attacked before they have time to untie them, and by that means rendered useless; a fatal example of which, and of neglecting the above precautions, I believe, will not be thought improper to be here inserted.

A Regiment of foot, consisting of above six hundred men, being ordered to march from one quarter to another, the commanding Officer imagined, from the distance of the enemy's frontier Garrisons, which was at least ten leagues, that he had nothing to apprehend, and therefore neglected the common precautions usually taken, in ordering his van-guard to examine all suspected places where horse might lye concealed. Besides, he took no care in the keeping up the divisions, but suffered the Regiment to run into a train of a mile long, in a very short time.

About half way lay a little wood, close to the road where the Regiment was to march, in which a famous Partisan, with eighty horse, lay concealed; which wood the van-guard passed without examining; and as soon as the center of the Regiment came opposite to the wood, the Partisan, with the eighty horse, rushed out upon them, and after killing about fifty men, and wounding as many more, the rest threw down their arms, and surrendered themselves prisoners; the men having the tent-poles fastened to their firelocks, could make little or no resistance. Besides, their marching in a straggling manner made the conquest easy, to which the surprise did not a little contribute, by making the number of the enemy appear infinitely more than they were.

The Officer who commanded the rear guard, hearing the fire in the front, and being about half a mile

in

in the rear, had time to put his men in order (which with his own, and those he had picked up, amounted to fifty) and stand upon his defence: and notwithstanding the disaster which happened to the Regiment, and some attempts to take him, he saved both himself and party, and retired back to the town in good order.

I would not have the world imagine, that I mention this affair, in order to reflect on the memory of the gentleman who commanded the Regiment; but by way of precaution to others: for he was known to be a brave man, and a very good Officer. His presuming too far on his safety, from the enemy being at such a distance, occasioned the infatuation with which he was then seized, (for I can call it nothing else, since it did not proceed from ignorance) the effect of which plunged both him and the Regiment into that misfortune. The disgrace of being surprised and taken by a handful of men, lay so near his heart, that it put an end to his life in a few years: and though the world forgive him, from the knowledge they had of his good qualities, yet he could never forgive himself.

The above misfortune is sufficient to convince us, that we ought not to slight or neglect our duty, particularly where the lives and safety of those under our command depend on our conduct.

To have the lives of men lost by our neglect, must touch our breasts with unspeakable grief, unless we are void of humanity.

But how can we answer it to our King and Country? or should they, like indulgent parents, pass it over, how can we reconcile it ourselves? The thoughts of it will sting us with remorse, and imbitter our lives to such a degree as to become a burthen to us. Such was the case, as I was informed, and which I am apt to believe, from the good character he bore, of the unfortunate gentleman before mentioned.

Had

Had the common precautions been obferved, without carrying it to a nicety, that misfortune could not have happened, though the enemy's horfe had been of an equal number with the Regiment; but by the forementioned neglect, that trifling, that inconfiderable body, made a bold pufh, and carried their point; which redounded very much to their honour, and the difgrace of the others.

We may draw advantages from the misfortunes of others, if we reflect juftly on them. It is with that view purely the above cafe was mentioned, and no other. Let us therefore avoid falling into the fame fnare, by acting with caution when we are intrufted with a command; and though we cannot be certain of fuccefs, with all our care and diligence, it is a great ftep towards it: for if we take proper meafures, our failing will then be attributed to the chance of war; and we may be unfortunate, though we do not deferve to be fo.

I will end thefe reflections, and this article, with a *French* maxim: *La meffiance eft la mere de la fureté.* Diftruft is the mother of fecurity.

ARTICLE IV.

When a Regiment, or Detachment, marches through an enclofed or woody country, the danger which they are to apprehend muft be from foot, and not horfe; and left a partifan party fhould efcape the difcovery of the van-guard, it would be proper to have fmall parties, commanded by Serjeants, marching on the flanks of the Battalion, with orders to examine all the hedges, ditches, and copfes which lie near the road, thofe being the places in which they generally conceal themfelves; and though the danger from fuch parties cannot be very confiderable, yet the neglecting them may occafion you the lofs of all your ftragglers, your baggage, and perhaps your rear-guard; confiderations,

derations, in my opinion, of too much weight to be slighted.

The parties on the flanks must by no means go too far from the Regiment, for fear of being cut off by the enemies slipping behind them; for which reason they must be very circumspect in their examining all suspected places, taking care to leave none behind them which they have not looked into, that their retreat may not be intercepted. Neither must they venture too rashly into a thicket or copse, lest they fall into their hands before they are aware of them, and by that means be taken prisoners, without being able to make a proper defence, till relieved by the Regiment.

Without these precautions, your parties may be taken within a hundred yards of the Battalion, in an enclosed or woody country, or have their retreat cut off by the enemy getting between them and the Regiment. When this is the case, they should give notice by the firing off of a piece or two, that relief may be sent them, and then make all the resistance possible till it comes, and not surrender upon any terms, but defend themselves to the last man.

When this happens, it would be very proper to sustain them immediately, by detaching the Grenadiers, or a platoon, or two, from that part of the Battalion which lies nearest; but with positive orders not to engage too rashly, for fear of being drawn into an ambuscade, and only endeavour to rescue their own men, without attempting any thing further, till they receive fresh orders from the commanding Officer how they are to proceed; for the usual decoy, by which people are drawn into an ambuscade, is, by laying of small parties at some distance from the place where the body lies concealed, which, at your approach, shew as if they were frightened, and retire with precipitation before you, in hopes to draw you into the trap; but when the Officers so detached have
effected

Chap. VIII. *Military Discipline.*

effected what they were sent for, that of disengaging their own men, they ought to pursue it no further, without fresh orders from the commanding Officer; otherwise, they may be engaged so far, that the commanding Officer shall find himself under a necessity to sustain them with the whole, and by that means be drawn into an engagement, before he has thoroughly considered whether it was proper or not.

Young Officers are but too apt to commit these mistakes, by exceeding their orders; being hurried on, by the heat and impetuosity of their temper, to do something that is great and noble, without considering the consequences that may attend it. I own it is an error on the right side; but it is still an error: for orders are, for the most part, positive, and leave us no room to act according to our inclination; a restraint that proves rather indulgent than hard in cases of danger, into which youth would precipitate themselves and others, were it not checked by the cool reason of men of experience. Let us therefore be subservient to the commands of our superiors, and submit to their judgment in all things relating to the service. We shall gain honour and reputation enough, if we adhere strictly to our orders; but disgrace may attend the exceeding of them, as well as the falling short; the one, however, is more excusable than the other, though the consequences may prove as fatal, since it proceeds from a mistaken zeal, but the other from the want of courage. To blame a man for want of courage, when nature has not bestowed it on him, is not only hard, but unjust; but a man that continues in the service, when he knows himself defective in that point, betrays both his King and Country, and therefore merits the severest punishment.

As soon as the advanced parties discover any troops, they are to acquaint the commanding Officer immediately with it, and whether they appeared to be horse or foot, what number, and which way they were

were marching; and so from time to time, in case they discover any thing further. Upon such a discovery, the advanced parties are to halt, and to remain there till they are ordered to retire, or forced to it by the enemy; in which case, they are to retire in a regular manner, and not with precipitation, lest they should intimidate the whole by a disorderly flight.

Upon such notice being given, the commanding Officer should immediately order the whole to halt, and prepare for action, and send the Major, or an Officer, that is well mounted, to reconnoitre them near, in order to discover what he can of their numbers and quality, and whether they appear to be friends or foes. But lest the closeness of the country will not permit him to do it without the danger of being intercepted in his return, small parties may be ordered to follow him at some distance, to secure his retreat; after the performance of which, they are to join the Regiment

The commanding Officer must direct his measures according to the report he shall receive from the Officer who was sent to reconnoitre; and if it is only judged to be a partisan party sculking about to pick up stragglers, or to take the baggage, he ought, no doubt, to order out a proper detachment to attack them; but with directions to the Officer who commands it, to be very cautious in the execution, by not pursuing too far, for fear of an ambuscade; and that if he should discover the body to be greater than what they apprehended, or find them too advantageously posted to be easily dislodged, to defer the execution till he acquaints him with it, and receives his further orders. Restrictions of this kind are not only proper, but absolutely necessary; without which, the party so detached may not be only lost, but the Regiment thereby involved in insuperable difficulties.

When the case happens, as above related, by their being too numerous, or too strongly posted to be attacked by the detachment so sent; the number of the enemy,

enemy, and their situation, should be thoroughly considered, and a proper disposition made to attack them to the most advantage, which must be done if they obstruct the march of the Regiment; but if they do not, I presume, the commanding Officer may desist from the attempt, if he finds much difficulty and danger in it, and pursue his march, taking care to secure the baggage (by marching it in the front, or on the flanks) and rear-guard, which he may do, by ordering it to be reinforced, and keeping near the rear of the Regiment.

This, however, must depend on his instructions, and by them alone he must be determined; it being impossible to say what should, or should not be done, without seeing them. For if his orders are only to march from such a place, to such a place, he is not to hazard his men in looking out for adventures; but to pursue his instructions to the best of his power, and attack those who shall oppose his march, provided they are not too powerful a body to encounter, or too advantageously posted to be forced; in which case, he may very justly retire to the place from whence he came, or into any other of safety that shall be nearer to him, acquainting the General, or Officer, from whom he received his orders, with the reasons for his so doing, and wait there for his further directions.

ARTICLE V.

The foregoing directions being more particularly calculated for the marching thorough an enclosed country, where the danger from foot was the chief thing to be apprehended, I shall, in the next place, lay down the necessary precautions for marching in an open, or champaign one.

In quitting the enclosed country, and entering into a plain or open one, the commanding Officer must take the necessary precautions against being surprised and attacked by horse; and though they may be discovered at a considerable distance, the quickness of their motions makes it proper that a disposition should be thought on, before there is a necessity to put it in practice

As the square is the principal figure into which a Regiment of foot can throw themselves against a considerable body of horse, they should be prepared to form it at the first order.

The method I propose for their marching in an open country, is as follows:

Upon their coming into the plain, or open country, the distance between the platoons should be no larger than what is required to form in, when ordered to wheel to the right or left; by observing of which, the Battalion may be formed in a moment, by one word of command, and ready to enter upon action, if required, in that position.

The parties on the flanks may join the Regiment, the van and rear-guards being sufficient.

In marching by platoons, the Regiment will take up the less ground, and become thereby more compact, and consequently not liable to a surprise; and, while they discover none of the enemy, they may pursue their march in that order; but, upon the appearance of any troops, or notice given them of their being near, the Battalion should then march in four grand divisions, the platoons being ordered to double up to the left, according to the number into which it was told off: for if it consisted of sixteen platoons, besides Grenadiers, then each grand division will consist of four, but if only into twelve, then each grand division will consist of three platoons.

The reason why I mention the Battalion being told off into sixteen platoons, or twelve, is, because the grand divisions are formed from one of those two in a moment, and from thence the square, without any alteration

alteration in the Officers or platoons in the forming of both; whereas a new divifion, both of the Officers and platoons, is required, fhould the Battalion be told off into thirteen, fourteen, or fifteen platoons, before either the grand divifions or fquare can be formed; which inconveniency, and lofs of time, is avoided, by dividing the Battalion into fixteen, or twelve platoons, exclufive of the Grenadiers, as may be feen by the plan of forming the fquare by divifions, in Article III. Chap. VII.

But as that only fhews how the fquare is formed by grand divifions, when drawn up in Battalion, I fhall fhew how it is to be formed from grand divifion upon the march. I fhall begin with fixteen platoons.

How a Battalion of fixteen platoons is to be formed into four grand divifions on the march, and then into the fquare.

The Battalion marching by platoons, the commanding Officer is to order the firft part of the *Affembly* to be beat, which is the ufual fignal for doubling; upon which the odd platoons, *viz.* the firft, third, fifth, &c. are to paffage in a direct line to the right: at the fame time, the even platoons, *viz.* the fecond, fourth, fixth, &c. are to move obliquely forwards, fo as to come up, upon the left of the platoons before them, by this movement forming fub-divifions. Upon the repetition of the fame fignal, the odd fub-divifions are to gain ground to the right, and the even ones, in an oblique direction, to the front, till they join the left flanks of the fub-divifions before them, forming by that means grand divifions. The Grenadiers in the front, and rear of the Battalion, are to keep oppofite to the center of the firft and fourth grand divifions.

By the above method, the Battalion may be formed into four grand divifions, even without halting, when the commanding Officer thinks proper; and from thence into the fquare, in the following manner:

As soon as the *Long-roll* beats, the sub-platoons of Grenadiers, in the front and rear of the Battalion, halt, then face to the right and left outwards, and march very briskly by files, till they are just clear of the flanks of the grand divisions, where the two platoons from the front, face to the rear, and join the flanks of the first grand division; the two from the rear, face to the front, and join the flanks of the fourth grand division.

The first grand division stands fast, and forms the front face. The other three grand divisions march briskly forwards, till they come to half distance, where the second and third, dividing in the center, wheel to the right and left outwards, by which they will form the right and left faces, thus: the two platoons on the right of each of those grand divisions wheel to the right, and form the right face; and the two platoons on the left of each, wheel to the left, and form the left face.

The fourth grand division continues marching, till its extremities join the forementioned faces.

The four platoons of Grenadiers remain dressed in a line, with the outside rank of the front, and rear faces.

The Ensigns who carry the Colours, together with the Drummers, form in the center of the square, facing to the front.

When the whole have taken up their proper ground, the drums cease, and the word of command, *Face square!* or *a Flam*, is given, at which the fourth or rear grand division, faces to the right-about, the Officers passing through the intervals of their platoons to the front, and the Serjeants to the rear.

The four platoons of Grenadiers, at the same time, face to the right-about, and, each of them wheeling upon its center, cover the angles of the square.

By keeping the grand divisions up to their proper distance, the square may be formed as quick, almost, as thought, without running the least hazard, though the enemy's horse should be within thirty paces: for which reason, the commanding Officer may continue his march in grand division till he finds they have made a disposition, and are actually on their march, to attack him on all sides, since the march will be quicker, and with less fatigue to the men, in marching by grand division than in the square. Neither do I think that a Regiment of foot can be under a necessity to throw themselves into a square for three or four squadrons of horse, since they may be kept at a distance, by marching the Grenadiers on the flanks, and to fire on them whenever they venture too near; but if that should not be sufficient, a few platoons from the Battalion will soon make them retire.

Upon discovering the enemy's horse, or notice given of their approach, the van and rear-guards, as also the baggage, should be ordered to keep near the Regiment; and if they find the number of the enemy considerable, and that the baggage cannot be preserved without running too great a hazard, it ought to be abandoned, and nothing thought of but the security of the Regiment; however, they need not give up the baggage, till they are under a necessity of doing it, nor defer it when they are.

When this is the case, the van and rear-guards should be divided into the several platoons, that they may not be exposed to the enemy, by being left out, when the square is formed.

If the enemy should not think proper to attack you, on their finding you drawn up in the square, and ready to receive them in too warm a manner, but retire to a greater distance, the commanding Officer may then pursue his march, either in the square, as has been shewn in Article IV. Chap. VII. or in grand divisions. If their distance is such, that he may march with safety, by grand divisions, the square is to be reduced into them, in the following manner.

The commanding Officer firſt gives the caution, *Take care to reduce the ſquare!* or *a Ruffle*: then the word of command, *Reduce the ſquare!* or *a Flam*: at which the ſecond and third grand diviſions, which formed the right and left faces, go to the right-about.

The four platoons of Grenadiers wheel briſkly on their center, and dreſs with the front rank of the front and rear faces.

The Enſigns go to the right-about, in order to march to their poſt, at the head of the third grand diviſion. The Drummers alſo go to the right-about, in order to march to their reſpective diviſions.

The whole having faced properly, the commanding Officer orders the *Long-roll*; at which the rear face, or fourth grand diviſion, marches to the rear.

The right and left hand faces, wheel to the right and left inwards by ſub-diviſions, till they join, and ſo form the ſecond and third grand diviſions.

The Enſigns fall into their poſts, and the Drummers march to the flanks of their diviſions.

The ſecond and third grand diviſions keep marching to the rear, till they have gained their proper diſtances from each other.

The two ſub-platoons of Grenadiers on the front, move forwards till they have got their proper diſtance from the front, or firſt grand diviſion; then facing inwards, march and join.

The two platoons of Grenadiers with the rear, move towards the rear, till they are at their proper diſtance; then, facing inwards, march and join.

This being done, a *Flam* is ordered, at which the ſecond, third, and fourth grand diviſions face to the right-about. The four platoons of Grenadiers, now making two, face to the right and left, to the front.

Officers and Serjeants of the fourth grand diviſion, paſs through the intervals of their platoons, to their former poſts.

When

When the Regiment is told off into twelve platoons, each grand division will consist of but three platoons; for which reason the square must be formed in this manner:

The first grand division forms the front face.

The second wheels to the right, and forms the right face.

The third marches on till they come opposite to the right flank of the second division, and then wheels to the left, and forms the left face.

The fourth division forms the rear face, as in the other.

In reducing the square, from hence, into grand divisions, the right and left faces are to wheel back, as they did up, entire; only the right face must not wheel back so quick as the left, that they may fall into their proper places, and avoid the confusion which their meeting, in wheeling back, might occasion.

This manner of forming the square, requires a little more time than the other; but in all other respects, it is quite as regular, by requiring no new telling off, or changing the Officers; whereas, should the second and third grand divisions divide in the center, and wheel to the right and left outwards, to form the right and left faces, as the other did, the center platoon of each of those grand divisions would be cut in two, and thereby occasion a new telling off of the right and left faces as soon as they were formed, which would not be quite so proper, if the enemy were very near, as that of wheeling up the grand divisions entire.

But as the difficulty seems to lie in the time which is required in the forming of the right and left faces by grand divisions entire; and that a greater will happen, by the dividing of them in the center, when they consist but of three platoons each, as has been observed; I shall offer therefore another way of doing it, as a medium between both, as thus:

The right face may be formed, by ordering the two platoons on the right of the second grand division, and the right platoon of the third, to wheel to the right to form that face.

The left face will consist of the left platoon of the second grand division, and the two on the left of the third; so that by ordering them to wheel to the left, at the same time that the others wheel to the right, the left face will be formed at the same time with the right; by which method the platoons, with their Officers, will be kept entire, and the square sooner formed than by grand divisions.

If the foregoing rules are observed, an Officer can never be so far surprized on his march, but that he will be ready to act upon the offensive, or defensive, which was all that I proposed to treat of in this Chapter; in the prosecuting of which, I have endeavoured to shew the duty of the whole, in such a manner, that every Officer may plainly discover his own; either in the proper directions for marching the whole, or a particular platoon or division, and passing a defile, or the command of the van, or rear-guard; or being ordered out with a detachment to sustain or relieve any of his own parties, or to attack a partisan party that shall be discovered on the march, with the precautions how they are to proceed in the execution, for fear of further danger than at first appeared, and how they are to act when it so happens; but, more particularly in the conduct required of a commanding Officer, for the security and preservation of the whole, both in an open and an enclosed country.

It is impossible to say in what manner he is to act when he meets with the enemy, without knowing their numbers, quality, and disposition. His own judgment and experience must direct him in taking proper measures; for without he has both, those who are under his command, at such a juncture, are much to be pitied, let his courage be ever so great.

CHAP.

CHAP. IX.

Consisting of general rules for Battalions of foot, when they are to engage in the line.

ARTICLE I.

THOUGH it may be supposed that all Colonels will keep their Regiments in such order, that they may be ready to march and enter upon action when commanded; however, it is absolutely necessary, that they make a thorough inspection into their mens arms and ammunition, the day before they expect an engagement, lest any thing should be wanting, when their service is required.

The commanding Officers should take particular care to have their Regiments as strong as possible on the day of action, by permitting no more men to be out of the ranks than what are absolutely necessary for the security of the baggage, or are commanded out upon duty.

As soon as the Battalions are formed, they should be told off into platoons, and the Officers appointed to them, as directed in Article I. Chap. VI.

When the Officers are posted to their platoons, they should view their mens arms and ammunition, and make a report of the same to the Officer commanding the Regiment; this inspection should not be omitted, lest several men, by having lost, or embezzled their ammunition, may be rendered useless for want of being supplied in due time, which they will be apt to conceal, for fear of being punished.

In

In marching up to attack the enemy, and during the action, a profound silence should be kept, that the commanding Officers may be distinctly heard in delivering their orders: neither are the Officers who command the platoons to use any more words than what relate to the performance of their duty.

The commanding Officer is to give the word of commands for all the movements which his Regiment is to perform, whether it is to advance, retire, or halt: but, lest he should not be distinctly heard by the whole, they are to regulate their motions by the several beatings of the drum; for which end, the Drummers in the center platoon are to be very attentive to the words of command, and to beat, on the delivering of them, according to the following directions.

When the Battalion is ordered to march forward, they are to beat a *March*; and when the word *Halt* is given, they are to cease.

When they are to retire, as soon as the Battalion has faced to the right-about, and the word *March* is given, the Drummers are to beat a *Retreat*; and not to cease till the Battalion is ordered to *Halt*.

The Drummers on the flanks are to govern themselves, both in their beatings, and in ceasing to beat, by those in the center: by which means, those men who could not hear the word of command, from their being at too great a distance from the center, or the noise of the drums, will know, by the different beatings, what they are to perform.

The Drummers in the center must be ready to beat a *Preparative* for the whole Battalion to make ready, if the commanding Officer thinks proper to have it performed in that manner.

ARTICLE II.

In marching up to attack the enemy, the line should move very slow, that the Battalions may be in order,

order, and the men not out of breath when they come to engage.

The commanding Officer of every Battalion should march up close to the enemy, before he suffers his men to give their fire; and if the enemy have not given theirs, he should prevent their doing it, by falling upon them, with the bayonets on the muzzles the instant he has fired, which may be done under the cover of the smoke, before they can perceive it: so that by the shock they will receive from your fire, by being close, and attacking them immediately with your bayonets, they may, in all probability, be beat with a very inconsiderable loss: but if you do not follow your fire that moment, but give them time to recover from the disorder yours may have put them into, the scene may change to your disadvantage. I therefore do not recommend this way of proceeding, but when the enemy are obstinate, and persevere in not giving theirs first; it being a received maxed, that those who preserve their fire the longest, will be sure to conquer: but if the method here proposed is duly executed, that maxim, I believe, will be found fallible. However, it should only be pursued in the case spoken of, as a proper expedient when you cannot draw the enemy's fire from them till you come up close; but if you can draw away their fire at some distance, without giving yours, and that the execution has not disordered the Battalion so much, but that it keeps moving on towards them, you may be sure of success; it being certain, that when troops see others advance, and going to pour in their fire amongst them, when theirs is gone, they will immediately give way, or, at least, it happens seldom otherwise. The point then to be aimed at is, that of receiving the enemy's fire first; but when both sides pursue the same maxim, in preserving their fire last, I do not know a more proper expedient than the one already mentioned: for when the fire is given near, there will not

be

be only a great many killed and wounded, but those who remain unhurt will be put into such disorder and confusion by it, that it will contribute to their being beat without much difficulty, if the blow is followed.

When it is apprehended that the enemy will persist in reserving their fire, the commanding Officers should prepare their men for it before they go on, and direct them how they are to give their fire, and in what manner they are to proceed afterwards, with the advantages that will be gained by the following of it, and that their own safety, as well as the destruction of the enemy, depends on the due execution.

ARTICLE III.

When any of the Battalions have forced those they attacked to give way, great care must be taken by the Officers to prevent their men from breaking after them; neither must they pursue them faster than the line advances: for if a Battalion advances out of the line it may be attacked on the flanks by the enemy's horse, who are frequently posted between the first and second lines for that purpose. The commanding Officers must therefore remain satisfied with the advantage of having obliged the enemy to give way, and not break the line, by advancing before it in the pursuit; but, in order to keep up the terror of the enemy, and to prevent their rallying, the Grenadiers may be ordered to advance twenty or thirty paces before the line, to fire upon them from time to time: and while the Grenadiers are thus employed, the commanding Officers should take great care to keep their Regiments in good order, that they may be ready to engage the second line of the enemy, which they may reasonably expect will come up to sustain those they had routed.

The Grenadiers being detached in the front only to prevent those who were routed from rallying, they must by no means advance too far from the line, lest they

they should be cut off from it by the enemy. They must therefore act with precaution; and as soon as they perceive the second line of the enemy, or a body of their troops, marching towards them, they are to quit the pursuit, and return to their Regiments, or halt till their own line comes up, if the enemy do not advance too fast upon them.

Unless these directions are punctually observed by every Battalion in the line, the advantage so gained may be snatched from them in a moment: for by pursuing the enemy too far, they may be surrounded by fresh troops, and cut to pieces, before the line can come up to their assistance. It is therefore the duty of every commanding Officer, to regulate his march according to the motions of the line, and not suffer themselves to be too much elated on the first success, lest it hurries them on too fast, without reflecting on the danger that may attend it: for which reason the whole line must act like one Battalion, both in advancing, attacking, and pursuing the enemy together. While they keep in a body, they can mutually assist one another; but if they should separate in pursuing those they beat, the enemy may destroy them one after another, with such an inconsiderable number of troops, that were they in a body, would fly at their appearance. The consequence therefore of separating during the action, is of such weight and moment, that by doing it, the enemy may not only re-establish their affairs in such a manner as to renew the action, but in all probability likewise gain a complete victory, if they make a proper use of the advantage so given; which we are always to suppose they will, and for that reason we ought not to give them an opportunity, by which they may have it in their power.

ARTICLE

ARTICLE IV.

Whereas it is ordered by the 13th article of the 14th section of the articles of war, 'that whoever, after victory, shall quit his commanding Officer or post, to plunder and pillage; every such offender, being duly convicted thereof, shall be reputed a disobeyer of military orders; and shall suffer death, or such other punishment, as by a general court-martial shall be inflicted on him.'

After what has been mentioned in the above article of war, relating to those who shall quit their posts to plunder and pillage, it appears almost unnecessary to enlarge on the necessity of its being duly observed, his Majesty's commands being sufficient to determine our obedience, without entering into the reasons for which they were given: however, it may not be altogether improper, for the information of those who have not had experience of the danger which generally attends the neglect of it, to shew that our own safety is thereby consulted and preserved.

Should the Soldiers be permitted to disperse, and run in search of plunder, before the enemy are entirely routed, and reduced beyond a possibility of forming themselves again into a body, they may renew the action, and fall upon them while they are employed in plundering the baggage; the consequence of which would not be only certain destruction to those who commit it, but perhaps prove so to the whole army likewise.

The history of all ages will furnish us with numerous examples of this kind: and a passage occurs to my memory, which I have read somewhere, that is very pat to the purpose.

A General of an army finding himself under very great difficulties, by being obliged to engage a superior force; and being apprehensive that the battle would

Chap. IX. *Military Discipline.* 161

would go against him, without some extraordinary means could be thought on to prevent it, at last came to the following resolution. As soon as he drew near the enemy, he ordered all the baggage of his army to be placed in their full view; after which he gave orders, that, upon the making of such a signal, they should make their retreat; which, however, should not be given till he found that the battle was likely to go against him; and as he conjectured, so it happened, from the superiority of the enemy: upon which he ordered the signal to be made, and his army retired in pretty good order, leaving the enemy masters of the field of battle, and all his baggage; the temptation of which, and their apprehending that they had nothing to fear from a beaten army, made them quit the pursuit, and fall a plundering. The General, finding that the bait which he had laid had taken effect, returned with his troops, and fell upon them in the height of their plundering, and by that means gained a complete victory.

Whether the above story is true or false, is of no great consequence. The carrying an air of probability in it, was sufficient to my purpose: and I believe, if the same stratagem was to be made use of, even in this age, it might have a very good effect in saving a great part of a broken army, by taking the enemy off from the pursuit; for such is the love of plunder in the private Soldiers, that were they not restrained from it by their Officers, no hazard would deter them from it.

As example is beyond precept, I beg leave to insert another of a more modern date, the truth of which I can assert.

In the year 1710, the late Earl of *Stanhope*, with a body of *English* troops, being attacked in *Brihuega*, by the whole *Spanish* army, consisting of above 20,000 men, under the conduct of the late King of *Spain*, and the Duke of *Vendosme*; to relieve which,

which, Marshal *Staremberg* marched with the remainder of his army, amounting to 11,000 men; but before he came up, the town was taken, and the Earl and his troops made prisoners; of which Marshal *Staremberg* could have no intelligence, till he had advanced too near to retire without fighting. The *Spanish* army looking upon their vast odds, and being elated with their success the night before in the taking of *Brihuega*, concluded on nothing less than the cutting of Marshal *Staremberg* and his little army to pieces. With this view, the duke of *Vendosme* detached a body of 3000 horse to fall upon the rear of Marshal *Staremberg*'s army, at the same time that he attacked him in front with the rest of the army; but the baggage of the Marshal's army being placed in the rear of his second line, saved him from the danger which threatened him there; for the 3000 horse, instead of doing their duty, carried off the baggage. As soon as the Duke of *Vendosme* had given a sufficient time for the 3000 horse to march round, as directed, he attacked the Marshal's army in front; but with so little success, notwithstanding his superior numbers, exclusive of the 3000 horse, that the Marshal did not only repulse the *Spanish* army in every attack they made, but beat them entirely out of the field of battle, and obliged the King of *Spain* and the duke of *Vendosme* to retire five miles that night: whereas, had the 3000 horse desisted from the plunder, and pursued their orders, in attacking the Marshal's army in the rear, when the Duke attacked it in front, few or none could have escaped.

As this short account is only to shew the ill consequences of suffering the Soldiers to quit their Officers or posts, to plunder or pillage, before the enemy are entirely beat; I hope, that the inserting of it here, will not be thought foreign to the affair in hand; but will rather serve as an example to deter others from committing the like error, and oblige every Officer

in

in his ſtation to exert himſelf on theſe occaſions, that the danger here complained of may be avoided.

ARTICLE V.

The commanding Officer and Major of each Regiment ſhould obſerve the diſpoſition of their own troops when they are drawn up in the line of battle, that they may know what Regiment ſuſtains theirs, or whom they are to ſuſtain, according as they are poſted in the firſt or ſecond line. They ſhould likewiſe endeavour to know their own troops from the enemy, either by their Colours or clothing, that in the confuſion, to which battles are generally ſubject, they may not miſtake the one for the other: for as the Battalions are obliged to ſeparate when a battle is fought in a cloſe or woody country, this miſtake is eaſily made, without the aforeſaid precaution, and therefore abſolutely neceſſary to prevent your deſtroying one another.

ARTICLE VI.

When the enemy act upon the defenſive, and only endeavour to maintain their poſt, if there are any houſes, hollow ways, ditches or hedges in their front, they commonly place men in them to annoy the line in their marching up to attack them. When this is the caſe, the Grenadiers ſhould be ordered to march thirty or forty paces before the line, either in ſingle Companies or joined, as the ſervice may require, in order to diſlodge thoſe advanced parties, that the line may not be ruffled, or interrupted, in their marching up to attack. As ſoon as the Grenadiers have performed that ſervice, they ſhould halt till the line comes up, and then join their Battalions.

ARTICLE VII.

When a Battalion is ordered to retire, or obliged to it without being beat or put into diforder; before the word of command, *To the right-about*, is given, the commanding Officer fhould acquaint the men with the reafon for fo doing, left they fhould apprehend the danger to be greater than it is, and thereby occafion their falling into confufion, inftead of making a regular and Soldier-like retreat.

The reafon for a Regiment's being ordered, or obliged to retire, may proceed from one of the following caufes.

When a Regiment has fuftained a great lofs in the firft line, it may be thought proper to relieve it by one from the fecond line; and therefore ordered to retire to make room for that coming up.

When a Regiment is expofed to the cannon of the enemy, they may be ordered to move to the right or left, or to retire to a place of more fafety.

Or when a Regiment is drawn up in fome advanced poft where it is likely to be overpowered by numbers, and cannot be eafily fuftained, they may be ordered to quit their poft and retire to one more fecure; but however juft the motives may be for their retiring, yet, without they receive orders from their Generals for it, they are not to quit their pofts, but muft maintain them to the utmoft of their power.

As every Battalion is to obferve the motions of the line, when they fee the Greateft part of it retire, they are to do the fame, without receiving a particular order from the General who commands the line for it; it being impoffible for him to fend fuch orders to every Regiment, in the heat of action; for which reafon, it is a fixed rule for every Battalion to act, as near as poffible, in concert with the whole, both in advancing, attacking, purfuing, or retiring together however, we are not to conclude from hence, if some of the Battalion should be ruffled in the attack and forced to give way, that the rest are obliged to follow their motions; neither are two or three Battalions to go on, when the rest retire. But whatever the motives may be for their retiring, whether those abovementioned or others, the commanding Officer should always acquaint the Regiment with it, and give his orders clear and distinct, without shewing any concern, otherwise the men may fall into confusion, for want of being apprised and duly prepared to perform what shall be ordered them; for if this precaution is omitted, and the words of command, for the changing of ground, or retiring, are given in a hurry, the men are apt to perform them in the same manner, and by that means occasion their falling into disorder, which, in the presence of the enemy, may produce dangerous consequences, by not only losing the reputation of the Regiment, but occasion several others to be seized with a pannic, and follow their example.

ACTICLE VIII.

When the first line is put into disorder by the enemy and forced to give way, the Battalions are to retire through the intervals left between those of the second line; but, to prevent their being too closely pursued, the Officers in the front should rally some of their boldest and most resolute men, and form them into small platoons, and fire upon them from time to time as they retire; which will not only oblige the enemy to advance with more deliberation, than they would do were there none to oppose them; but it will likewise do them considerable damage, and give their own Battalions an opportunity of making a more safe and orderly retreat.

The Officers who are posted in the rear of the Battalion, as also the Major and Adjutant, must prevent their men from retiring too fast, and endeavour to keep them in a body, and from mixing with those of another Regiment, and to march them in the best manner they can through the intervals of the second line, which when they have passed, they are to halt and form them again into Battalion, with the utmost expedition, in order to march up and sustain the second line.

On these occasions, the danger which the second line runs of being broke by the first, is very great; for when the first line is put into disorder, instead of retiring through the intervals of the second line, they frequently run directly into the Battalions, and either carry them off with them, or put them into such confusion, that the enemy are upon them before they have time to repair the disorder: for which reason, the Officers of the first line must use their utmost diligence and care to prevent their men from committing this error, by observing the above directions; without which, their own preservation is not only obstructed, but the loss of the whole must inevitably follow.

ARTICLE IX.

As the case above-mentioned, that of the second line being broke by the first when they are forced to give way, is very common, I shall therefore offer the following expedient, in order to prevent the evil complained of, and what, in my opinion, will be very conducive thereto.

As soon as the first line gives way, the Grenadiers of the second should be ordered to advance twenty paces in the front, and directly opposite to the center of their own Regiments, in order to keep the men of the first line clear of them, and oblige them to retire through the intervals, or compel them to it by force; but that should be used with moderation:

However, in cases of danger, as this is, it is better that a few should suffer than the whole.

By advancing the Grenadiers into the front of each Battalion, the intervals of the second line will be considerably enlarged by it, and consequently opens a greater passage for those of the first line to retire through; by which, the danger of the second line's being put into disorder by the first, will be less; which consideration alone appears to me of sufficient weight for its being done.

But I am likewise of opinion, that it will not only prevent the second line from being thrown into confusion, but prove, in some measure, a security to the first, by putting a check to the enemy in their pursuit, and either oblige them to halt, or advance slower, when they perceive the Grenadiers of the second advancing in order, and ready to oppose them: but whether it has this effect or not, the other advantages proposed by it, that of securing the second line from being broke or put into disorder, and the enlarging of the intervals for the first line to march through, are sufficient motives for doing it.

The design of advancing the Grenadiers, being only to prevent the first line from mixing with the second, and to stop the pursuit of the enemy, by shewing themselves to them, in order to facilitate the retreat of the men of the first line, and to prevent a further execution on them, they must therefore by no means engage; but as soon as the men of the first line have gained the intervals, they are to join their Regiments.

This method, of ordering the Grenadiers of the second line to advance upon the first line's giving way, being purely a notion of my own, neither having seen nor heard that it was ever put in practice, I shall not lay it down as a fixed rule; therefore every body is at liberty to reject it at pleasure.

ARTICLE X.

Upon the first line giving way, the second should march up briskly to their relief, and attack the enemy before they have time to repair the disorder which both the action and the pursuit must, of course, have thrown them into; by doing which, they may, with ease, retrieve what was lost; and, in all probability, do such execution, that those troops, who were before victorious, may be rendered incapable of any further service that day: for we may reasonably suppose, that they will be considerably diminished, and put into disorder, by their action with the first line; and as their first fire is spent, which is the one that does the most execution, the others, from the too great hurry in loading, are of no great consequence; at least in comparison with the first: whereas, on the other hand, the second line being entirely fresh, in good order, and their arms well loaded, and the edge of the enemy's vigour somewhat blunted, we may justly conclude, that they will not be able to resist the attack of the second line, if performed like brave and resolute Soldiers: so that by the loss which they sustained from your first line, and the immediate attack of the second, the above supposition, that of rendering those troops unserviceable, appears both a reasonable and a well-grounded assertion.

But the greatest difficulty which we have to struggle with on these occasions, does not proceed so much from a real as an imaginary danger: for when the men of the second line perceive those of the first give way, they are apt to form to themselves vast ideas of the enemy, which, by working strongly on the imagination, become so terrible, that, by the time they approach near, they frequently betake themselves to a shameful flight, or make but a weak resistance: so that by being seized with a false fear, they do not

only

only lose the opportunity, which fortune throws in their way, of gaining an advantage over the enemy; but quit even the means by which their own lives might be saved: whereas, if they would but give themselves time to consider, their own reason must tell them, that fresh troops will always prove too powerful for those who have been already engaged; and that, if they acted as became Soldiers, they could not fail of success in the present case. But we cannot expect that the majority of the private men will reason in this manner, since their reflections proceed rather from what is conveyed to them by the eye, than from the understanding: however, I must do the common Soldiers of our kingdom the justice to say, that they are as seldom guilty of the failing here complained of, as those of any other nation in *Europe:* and that whenever it does happen, it proceeds oftener from the want of conduct in the Officers, than courage in the men: for the *English* are naturally active, strong, bold, and enterprising; always ready to go on to action; but impatient when delayed or kept back from it. I should therefore think it adviseable, in the case we speak of, for the second line to move on briskly upon the first's giving way; by which means, they will not only save the lives of a great many men of the first line, but may fall upon the enemy before they can have time to put their Battalions in order, which is a great point towards rendering their success both certain and easy: but if the second line should halt, or move but slowly on, it will give the enemy time to do great execution on the men of the first, the sight of which may strike those of the second line with a pannic; whereas, the carrying them on briskly will divert their eyes from the sight of the slain, or at least not suffer them to remain so long as to prove of any ill consequence, or abate their courage.

As there is not any one part of the service more difficult to manage, than what is above treated of;

or

or where the courage and conduct both of the Officers and Soldiers are put to a greater trial; I was therefore induced to be the more particular on that head, in order to remove the wrong impressions which those of the second line generally receive upon their seeing the first line give way; and likewise to shew how easy it is to repair that loss, and gain immortal honour, if they act as they ought to do; which design, (though I am afraid very unskilfully executed) I hope will excuse the length.

ARTICLE XI.

It being a general remark, that the private Soldiers, when they are to go upon action, form their notions of the danger from the outward appearance of their Officers; and according to their looks apprehend the undertaking to be more or less difficult: (for when they perceive their Officers dejected, or thoughtful, they are apt to conclude the affair desperate:) in order therefore to dissipate their fears, and fortify their courage, the Officers should assume a serene and chearful air; and in delivering their orders to, and in their common discourse with the men, they should address themselves to them in an affable and affectionate manner.

When the private Soldiers have an opinion of the military capacity of their Officers, or have had experience of their courage and conduct, the above method will effectually prevail, and create in them such an opinion of their own superiority over the enemy, that they will look upon them with contempt, and conclude them, in a manner, beat before they begin the action. When such a spirit is once raised in the men, they seldom or ever fail of success: but when Officers have not had experience of the service, or neglect the means by which they might attain to it, (of which the private men are strict observers, and

from

from thence form their judgment of them) the cafe will not hold: for unlefs the men have an opinion of their conduct, as well as their courage, they will not be able to influence in the manner above fpoken of.

ARTICLE XII.

When the line is marching to attack the enemy, and that a Battalion is fo ftraitened for want of room that they cannot march the whole in front, by the others preffing too much upon them, or that the ground will not admit of it, an entire platoon fhould be ordered to fall back and march in the rear of the Regiment, till the interval will allow of its moving up; and in cafe the Battalion fhould engage before the platoon can move up, it will not only prove a good referve, but be of confiderable fervice in keeping the reft up to their duty.

ARTICLE XIII.

Having treated at large, in the fecond chapter of this book, how a Battalion of foot, when detached, is to proceed when attacked by horfe; and as thofe in the line may be liable to it during fome part of the action, the commanding Officer of every Battalion fhould be prepared to receive them according to the method laid down in the faid chapter; with this difference, that when they are attacked by horfe in the line, or in brigade, they are not then to act feparately, but in conjunction with one another. The method of acting in this cafe, muft depend on the manner you are attacked; for if they endeavour to fall on the flank of the line, by its laying open to them, a Battalion or more muft be wheeled back to fecure it. But as this would carry me into a higher fcene than what I propofed, or am qualified for, I have therefore endeavoured to avoid it as much as I poffibly could;

could; and have only entered so far, as not to leave the parts treated on dark or obscure, that the young Officers, for whom it is writ, may have a clear and perfect notion of the several parts of their duty, in order to execute what shall be commanded them by their superiors; without which, it was impossible for me to make them comprehend it.

ARTICLE XIV.

I intended to have ended this chapter with the above article; but as the infantry of *Holland* begin the action, when they engage in an open or champaign Country, with firing by platoons as they advance upon the enemy; I believe it will not be improper to set down their method of performing it, with the advantages and disadvantages that may arise from that way of proceeding.

Upon the line's approaching so near the enemy that they can do execution on them with their fire, which I suppose to be about sixty paces, all the Battalions make ready, and march on with recovered arms; a little after which, the platoon on the right of each Battalion advances forward, till the rear rank comes even with the front rank of the Regiment, then halt, kneel, lock, present, and fire. As soon as the platoons on the right have fired, those on the left advance and do the same; and so from the right and left of each Battalion till they finish in the center, by making every platoon fire once. This way is called, Alternate Firing.

In performing the alternate firing, the whole line moves as slow as foot can fall, that the Battalions may not fall into any disorder, by the platoons advancing to fire; as also to give the men time to load their arms before they approach too near the enemy; which when they have done, the line then moves on
brisk-

Chap. IX. *Military Discipline.* 173

briskly; and when they come up close to the enemy, they give them their whole fire, as the *English* do.

By this way of proceeding, the enemy must receive two fires; but it has sometimes happened, that, by their alternate firing only, they have done such execution, that the enemy did not wait for their second fire; or if they did, they were too much weakened and disordered by it to make their resistance prove of any great consequence; by which means, they have often beat those they attacked with a very inconsiderable loss to themselves.

When a body of foot can be brought to perform the alternate firing in a cool and regular manner, it must be owned, that they will have a great advantage over those they attack, if they do not act in the same manner: but unless the Officers have a thorough knowledge of the service, or at least those who command the platoons, and the private men very exact in obeying the words of command, and expert in loading after they have fired, it will be dangerous to attempt it.

For should the enemy, by a quick and sudden motion, advance upon them before they have all loaded (which would be right for them so to do) those who have not loaded will be apt to give way, from a notion of their being then defenceless; the consequence of which may throw a pannick into the whole, and involve them in the same misfortune: therefore, unless it is managed with great conduct, it may very easily turn to their disadvantage.

But the *Dutch*, from the excellency of their discipline, which they strictly adhere to, have surmounted these difficulties: for as the selling and buying of commissions is a traffic (and I believe the only one) unknown, or at least not allowed of, in the Republic; their Officers are generally promoted by their service; by which means, the majority of

them

them are men of experience, and therefore may undertake it without running much hazard: for as they are judges of the service, they will easily perceive by the motions of the enemy, when they should continue or desist from the alternate firing, and, consequently, can avoid the danger above-mentioned, that of being attacked by the enemy before all their men have loaded.

We have a common notion, that this *Sang froid*, or obedient quality in the *Dutch*, is owing chiefly to nature, by their having a greater proportion of phlegm in their constitution than the *English*, by which their minds are not so soon agitated as ours. But I look upon this way of reasoning, to be rather a plausible excuse for our own neglect, in not bringing our men to the same perfection of discipline, than the production of any natural cause in the *Dutch*. But, allowing that nature does contribute something towards it, yet, it is evident, art has the greatest share, since their troops are generally composed of different nations.

The *French* form the same notion of us, as we do of the *Dutch*; but I am surprized that nation should still entertain such an opinion, considering how often we have given sufficient reasons to convince them that we do not want a due proportion of fire: and I hope we shall never have an occasion of being convinced, by such powerful arguments, that the *Dutch* are not defective in that point.

It is allowed by all nations, that the *English* possess courage in an eminent degree; but, at the same time, they accuse us of the want of patience, and consequently that which it produces, obedience; so that by our being defective in the latter, the great advantages which might be reaped from the former are often lost, or at least fall short of what might be justly expected from it. This accusation is something severe,

severe, since by it they deprive us of an essential quality (and, I was going to say, the most essential one) towards the forming of a Soldier, without which, no man can be justly stiled a complete one. They might say as well, that our courage is the effect of passion, and not reason: and, indeed, as it generally appears to them, they have too much colour for the assertion, by our neglect of discipline; and from thence conclude, that the *English* cannot be brought into it: but they are certainly mistaken in that, since none are more capable of instruction than the *English*; and when proper means are used, neither patience, nor obedience, are wanting in them.

CHAP. X.

Consisting of the duty of the Infantry in Garrison.

ARTICLE I.

HAVING treated in Chap. VIII. of the necessary precautions for marching a Regiment of foot, &c. I shall proceed to the several parts of the duty which is to be performed before they arrive at, and during their stay in, garrison; according to the modern practice of the garrisons abroad.

The day before the Regiment is to arrive, the commanding Officer should send the regimental Quartermaster, or a proper Officer, with an exact copy of his route or orders to the Governor, or Commandant of the garrison, to apprise him of their coming, that he may give the necessary orders for their being quartered.

When the said orders are given, the Quarter-master is to go along with the Barrack-master, or person appointed for that purpose, to take possession of the Caserns, or Barracks, alloted for the Regiment, and to examine nicely into the condition of the said quarters, in order to have them repaired, if they want it, before the Regiment arrives.

He is likewise to take an exact inventory of all the furniture, as beds, tables, stools, and the utensils for the dressing of victuals, &c. and what condition they are in, that it may be inserted in the receipt which he is to give to the person who delivers them.

If there is no conveniency for the lodging the Officers with the men, they should be billeted on the houses

houses which lie most contiguous to the caserns, or Barracks of the Regiment, that they may be ready, on all occasions, to join their men.

He is to make an exact division of the caserns, or rooms, according to the number of the Companies in the Regiment; placing the number designed for each, contiguous to one another; after which they are to be put into lots, to be drawn for when the Regiment arrives.

He is likewise to receive the fire and candle allotted for the Regiment, and to deliver it to the men as soon as they are quartered.

ARTICLE II.

If the troops in the garrison are so numerous, that several Regiments must be quartered on the inhabitants, for want of caserns or barracks to contain them; the town should, in that case, be divided into as many parts as there are Regiments to be so quartered; that the Officers and Soldiers of each Regiment may have a distinct part to themselves, in order to their being drawn together without loss of time, in case of an alarm; as also to avoid disputes which might arise betwixt Regiment and Regiment, by having them intermixed with one another. It will likewise be a great ease to the Officers in the visiting of their men's quarters; and to the Serjeants and Corporals in delivering of orders to the men, and in assembling of them for the parade, exercise, or the viewing of their arms and accoutrements.

This way of quartering, is called, in the military phrase, cantoning of a town.

ARTICLE III.

When the Regiment comes near the garrison, the commanding Officer should send an Officer, to acquaint the Governor, that he shall arrive at such a time, and to desire he will send orders to the Officer of the port-guard to permit the Regiment to march into the town; without which precaution they will be kept without the Barrier, till the Officer of the guard sends and receives the Governor's orders for their admittance.

On this notice, the Governor commonly sends the Town-Major, or one of the Town-Adjutants, with orders for their entrance, and to conduct the Regiment to the Grand-Parade.

The Quarter-master, or Officer who was sent to prepare the quarters, should meet the Regiment at some distance from the town, to acquaint the Colonel with his proceedings; as also to find out some convenient place near the town for the Regiment to draw up before they enter, and to conduct them to it. This small halt is only to give the men time to roll their cravats, cock their hats, and put themselves in the best order they can, that they may appear in a decent and Soldier-like manner upon their entering the town.

While the men are putting themselves in order, the Officers billets may be distributed amongst them, which is commonly done by lot, to avoid shewing any partiality; that is, those of each rank are to draw with one another. The reason why I think that the Officers billets should be distributed before they march into the town, is, that they may send their baggage directly to their quarters, and not lose time, by deferring of it till they come to the Parade; but this may, or may not be done, as the Colonel shall think proper.

As soon as the men have put themselves in order, and the files are completed, they may then pursue their march; and when they come near the Barrier, all the Officers, except the Colonel, Major, and Adjutant, must dismount, and march at the head of

their

their divisions with their Espontons in their hands, and the Ensigns the Colours.

It is a standing rule, that when the Officers are ordered to dismount, and the Ensigns to take the Colours, the Drummers are to fall into their divisions, and beat a *March*, unless the service may require their being silent; in which case, orders are given accordingly.

The Colonel, or commanding Officer, marches always an horseback at the head of the Regiment; but when the Governor, or superior Officer, is to see the Regiment march by, as soon as he has notice of the Governor's approach, he is to dismount, and march with his Esponton in his hand, and pay him the compliment due to his character in the army.

ARTICLE IV.

As soon as the Town-Major, or the Officer appointed to act for him, has notice from the Sentinels that the Regiment is in view, he should take a Serjeant and a file of men, and go to the outermost Barrier, and order one of the draw-bridges to be drawn up after him, till he has examined the original orders or route of the Regiment, lest the enemy, by having notice of the march of the Regiment, should, under that pretence, endeavour to surprise the town.

The Town-Major, and the party from the guard, are to remain within the Barrier; and when the Regiment approaches near it, he is to order the gate to be shut; upon which the Colonel shall halt the Regiment, and send the Major with the original order for his marching to that garrison, to be perused by the Town-Major, who is to receive it over the Barrier; and when he finds it authentick, and has discovered the Regiment to be friends, he then orders the gate of the Barrier to be opened, the draw-bridge to be let down, and the Regiment to march in.

When the Colonel comes up to the Barrier, the Town-Major is to return him the route or order, and then conduct the Regiment to the Grand-Parade, where they are to draw up in Battalion; after which, the Colonel, attended by the Town-Major, is to wait upon the Governor, to whom he is to deliver the original order for his marching to that garrison, and, at the same time, acquaint him with the state of the Regiment, and deliver him a return of it in writing.

During the time the Colonel is at the Governor's, the Regiment may order their arms; but no man must be suffered to stir out of the ranks.

After the Governor has perused the route, and the return of the Regiment, and asked such further questions concerning it, as he shall think proper, he then orders the Town-Major to wait upon the Colonel back to the Regiment, and to read the general orders of the garrison to the Officers and Soldiers, that they may not commit a crime through ignorance; and then to conduct the Regiment to the Alarum-post assigned them, and afterwards to their caserns, barracks, or cantonment, where he is to dismiss them.

Upon the Colonel's returning from the Governor's, he is to be received with shouldered arms, and all the Officers at their posts with their Espontons in their hands; which ceremony is always due to a Colonel from his own Regiment.

It is a standing rule, that whenever the compliment of rested arms is paid to any person, or from one body of troops to another, it is always to be done from shouldered arms.

The Colonel then orders the men to shoulder, and the Battalion to be told off into three grand divisions; after which, the flank grand divisions to be wheeled to the right and left inwards, and the Grenadiers to draw up opposite to, and facing the center grand division. This being done, he commands them to keep a profound silence; and the Town-Major being placed

placed in the center of the fquare, reads the general orders of the garrifon, and delivers an extract of the fame in writing to the Major, that each Officer may have a copy of them.

After this they are to be reduced into Battalion; then the Town-Major conducts them to their Regimental Parade, from thence to their Alarum-Poft, and afterwards to the Quarters affigned them, where, (after a proper guard is appointed for the fecurity of the quarters, and for relieving the Colonel's Sentry, and the colours lodged) they are to be difmiffed.

It is a general rule for every Regiment to furnifh their proportion of men to the Town-guard the day after they arrive; but when that can be difpenfed with, it would be proper to give them that day to clean their arms and accoutrements: however, this depends on the Governor, whether he will or will not allow of it.

CHAP. XI.

The usual Guards in a garrison, with directions for forming the parade.

ARTICLE IV.

THE Guards are composed of a Detachment of men taken from each Regiment, who are to mount daily for the security of the place.

The number of men who mount daily, must depend on the strength of the troops in garrison, and the number of posts to be guarded; according to which the duty is calculated: however, it is a fixed maxim in most of the garrisons abroad, to calculate the duty in such a manner, that the Soldiers shall mount guard every third day; and though the troops should be very numerous, they never suffer them to be above three days off, and the fourth on duty. This is done by mounting of more guards than usual, or by adding to the number of each guard.

The main-guard is generally composed of a Captain, two Subalterns, two Serjeants, two Corporals, two Drummers, and forty-eight, or fifty private Soldiers.

Each post has a Subaltern, Serjeant, Corporal, Drummer, and twenty-five or thirty private Soldiers.

Where there are Outworks of consequence, such as redoubts, detached Bastions, &c. Officers guards mount commonly in them; but more particularly so,
in

in frontier garrisons; otherwise Serjeants guards may be sufficient.

The citadel has generally a garrison for its own security, composed of an entire Regiment or more; in which case, those troops do the duty of the citadel only, without interfering with that of the town, or furnishing any men for its guard.

In frontier towns, they commonly mount another guard, called the *Reserve*, being much in the same nature as a Picquet-guard in camp, which is to be ready to march whenever the Governor shall order them.

The Reserve Guard consists commonly of a Captain, two Subalterns, two Serjeants, two Corporals, two Drummers, and fifty men, or a greater number if requisite.

The time of mounting the town guards, is commonly at eight, nine or ten in the morning: however, that depends on the Governor.

ARTICLE II.

The men who mount the guard, are to assemble at their Captains quarters, to be viewed by them, or their Subalterns, to see that their arms, ammunition and accoutrements are in good order, and that they are clean and well dressed; after which, the orderly Corporals are to march them to the Regimental-parade, and deliver them over to the Adjutant, who is to draw them up according to the seniority of their Companies, and to see that each Company has furnished the number appointed. He must likewise size them as soon as they are formed, which should never be omitted, even in a detachment of twelve men, since it will add vastly to their appearance.

As soon as the Detachments from the several Companies are drawn up, the men sized, and the files completed, the Adjutant and the Officers of the Regiment,

ment, who mount that day, are to view the mens arms, ammunition, clothes and accoutrements, that there may be nothing wanting when they come to the Grand-parade.

By this infpection, the Adjutant will know if the Officers of the feveral Companies have complied with their duty, in viewing their men before they were brought to the Regimental-parade; and when he finds that they have been remifs, by the men not being clean and well dreffed, or any thing wanting, he is to acquaint the Colonel with it, that they may be reprimanded for their neglect.

In fome garrifons it is the cuftom to have the men, who mount, on their Regimental-parade two hours before the beating of the *Affembly*; in order to be exercifed a-part by their own Adjutants. In others, they are to be at their Regimental-parade only half an hour before the *Affembly*; in which cafe, the whole are exercifed together on the Grand-parade by the eldeft Officer who mounts the guard.

It may be done in this manner, when the garrifon is compofed of troops of one nation, and that the number who mount do not exceed 600 men; but when it is compofed of different nations, or that the numbers are too great to be exercifed together, the former method muft be followed, that of the feveral corps being exercifed a-part, on their Regimental-parade, by their own Adjutants.

ARTICLE III.

The Drum-Major, with all the Drummers of that Regiment which gives a Captain to the main-guard, are to beat all the beatings of the mounting of the guard.

They are all to parade at the head of the main-guard half an hour before the time of beating; and when the hour appointed is come, the Drum Major

is

is to form them into ranks, and placing himself at the head of them, orders them to beat the *Assembly*, which they are to do quite round the Grand-parade, and back to the main-guard; after which, the Drummers, who mount, separate, and march to their Regimental-parade, beating the *Assembly* the whole way; but the Drum Major, with the others who do not mount, beat back to the Grand-parade, where they are to remain till the guards are marched off.

As soon as the Drummers return to their Regimental-parade, the Officers, who mount the guard, are to march their men to the Grand-parade, where they are to draw up, not by seniority of Regiments, but according to lot, which the Adjutants are to draw for, before the Detachments arrive, that they may shew them where they are to draw up.

The main-guard is always composed of the Officers and Soldiers of one Regiment, each taking it in its turn, and beginning with the eldest.

The Regiment which mounts the main-guard, draws up on the right of the parade: the Detachments of the other Regiments having no fixed post, are to draw up according to the lot drawn for.

The reason why they draw for their posts on the parade, appears to me as follows.

Should the Regiments have a fixed post on the parade, by drawing up constantly by seniority of Regiments, the men could then know what guard they were to mount, and by that means have it in their power to carry on a treacherous correspondence with the enemy, for the delivering up of a port or outwork of consequence; but as their posts are drawn for daily, they cannot know where their lot will fall, the uncertainty of which will keep them out of temptation, and effectually prevent any design of that nature; for which reason, neither Officer nor Soldier is allowed to change his guard with another: which maxim, of the Regiments drawing for their posts on the parade, and

the

the not suffering of the Officers and Soldiers to change their guards with one another, I presume, is founded on the sad experience which a contrary proceeding had produced.

When the Detachments from the several Regiments are drawn up on the Grand-parade, the Serjeants who mount are to form themselves in a rank entire opposite to their own Detachments, facing the men, and four paces advanced, in order to be posted to the guards, as the Town-Major tells them off.

ARTICLE IV.

The Town-Major begins on the right to form the guards; and as each is told off, he posts the Serjeants to them, taking them first from the right. As each guard is told off, and the Serjeant or Serjeants posted to them, he commands the men to order their arms, in which position they are to remain till ordered to shoulder.

As the Serjeants are posted to the guards, they are to fall into the intervals on the left of the front rank of their guards; but where there are two to one guard, the other is to fall into the rear of it, in order to prevent the men from leaving their ranks, or changing their guards, which those in the front are likewise to have an eye to.

The Officers are to draw for their guards, those who mount the main-guard and reserve excepted; after which they are to give in their names, with the Regiments they belong to, and the guard they have drawn, to the Town-Major, who enters them in a book. The names of the Officers who mount the main-guard are to be given by their Adjutant, to be entered with the rest.

The reserve being a distinct duty from that of the town-guards, it is always to be drawn up on the left of the parade, or formed after the rest are marched

off

off. It is compofed of an equal number of men from each Regiment, and a roll of duty kept apart for the Officers.

When a Guard of horfe mounts, it is to be drawn up on the right of the main guard, leaving an interval between them.

The guards being formed and told off, the Drum-Major, with all the Drummers, who affembled at the main-guard, are to beat the *Troop*, along the head of the guards, beginning upon the right, and marching from thence to the left, and back again to the right, where they are to finifh.

When the *Troop* beats, the Officers are to take their Efpontons in their hands, and place themfelves at the head of their guards, facing the men; the Serjeants are, at the fame time, to fall into their divifions, and the Drummers to place themfelves between the firft and fecond ranks of their guards.

After the *Troop* is finifhed, and the Officers have taken their pofts, the Town-Major orders his orderly Drummer to beat a *Ruffle*, by way of *Preparative*; then three *Flams*; the firft to reft; the fecond to fhoulder; and at the third, the Officers come to the right-about to their proper front, at three motions, and immediately drefs in a line.

When the whole parade is to be exercifed together, the eldeft Officer, who mounts, is to give the words of command, and to proceed in the fame manner as is directed for the exercifing of a Battalion, but to go no further than the manual exercife. If the eldeft Officer has not a voice ftrong enough, or any other impediment, upon his fpeaking to the Town-Major, I prefume, he will excufe him, and order an Adjutant of one of the Regiments to do it, or order it to be done by beat of Drum.

As foon as the exercife is over, the Town-Major orders the guards to march off, one after another, beginning with the right; but no guard is to move till
he

he says to every Officer who commands a guard, *March!* The reason for this is, that he may have time to view every guard distinctly, to see if they are in order, and that they have their complement of men.

When the main-guard is posted on the parade, all the others march off before it.

As soon as the guards of foot are marched off from the parade, the guard of horse is then to march; but it is not to move till all the foot are marched.

This must proceed from an old custom, that of the foot having the rank in garrison, and the horse in the field; by which the youngest Captain of foot commanded all the Captains of horse, while in garrison; and the youngest Captain of horse commanded all the Captains of foot in the field, without any regard to the dates of their commissions. But that custom is now abolished, and every Officer commands according to seniority, whether of horse, foot, or dragoons; however, that custom, of the foot guards marching off first, still subsists; because they are not only more numerous, but the immediate security of the place is committed to their charge: whereas the guard of horse is only employed to patrole during the night in the streets, and to reconnoitre the avenues leading to the town at the opening of the gates, to discover if the troops of the enemy are near, that the town may not be surprised.

When the Drummers of the main-guard, or that which marches off first, beat a *March*, all the Drummers of the other guards are to do the same.

As soon as the guards are marched off, the Town-Major should wait upon the Governor, or Commandant of the garrison, and acquaint him with it, and know if he has any commands for the garrison; after which, he returns to the parade and dismisses the Adjutants, or sends them notice by an orderly Serjeant, that there are no further commands for them

Chap. XI. *Military Dicifpline.* 189

at that time; till which they are not to leave the parade.

The orderly Serjeant and Corporal of each Company are to attend the parade every morning, and to remain there till the Adjutants are difmiffed, that if any orders are to be delivered, they may be ready to receive them.

In all the *French* garrifons, as foon as the parade is formed, and the guards ready to march, the Town-Major acquaints the Governor, or Commandant, with it, who is obliged to come and fee the guards march off; but in the garrifons of *Holland*, the Governors lie under no fuch injunction, though they frequently do it.

In garrifons which are remote from the enemy, the orders are generally delivered out immediately after the guards are marched off from the parade; but in frontier towns they are never given out to the Adjutants till the gates are fhut.

When the King, a Prince of the Blood, the Captain-General, or a perfon of authority who is entitled to a guard, comes into a garrifon, the eldeft Regiment is always to mount a proper guard on him during his ftay there, without rolling with the others, or having any allowance for it in the town-duty; being to furnifh the fame number of men for the parade, referve, and detachments as before; this is the cuftom abroad; but I prefume it is meant only when fuch great perfons ftay a night or two; for fhould they continue any confiderable time, the duty would fall too hard upon one Regiment; which rule, in my opinion, may be very juftly broke through by either allowing them for it in the other duties, or by ordering each Regiment to take that guard in its turn; the latter of which appears the moft equitable, becaufe the town-guards and detachments are much more fatiguing than thofe mounted on great perfons.

ARTICLE V.

The Officers who are to be relieved, are to order their men to stand to their arms as soon as they hear the Drum of those who are coming to relieve them; and when they come in sight, they are to order their men to shoulder their arms. When the new guard approaches very near, the Officer of the old one orders his men to rest their firelocks, and the Drummers to beat a *March*.

The Officer who comes to relieve, is to draw up his guard opposite to the old one, in the same manner that they are, whether six or three deep, or in a rank entire; and when the ranks are dressed, he then orders his men to rest their firelocks; in which position both guards are to remain till those who mount in the outworks have marched past them, which they are to do between the two guards, provided they are the port-guards, otherwise they cannot interfere with those sent to the outworks. After this, the Officers advance towards one another, paying the usual ceremony with their hats, and the Officer who is to be relieved, delivers all the orders relating to the guard to the Officer who comes to relieve him, acquainting him with the number of Sentries by day and by night, what patroles, *&c*. The Serjeants and Corporals of the old guard deliver their orders to those of the new at the same time; and when that is over, the men of both guards should be ordered to shoulder; the Corporal of the new guard is to number his men, and to draw out the number of Sentries who are then to be posted, forming them into ranks, and, being conducted by the Corporal of the old guard, march with them to relieve the Sentries. They go first to the Sentry who is posted the furthest from the guard, and relieve him, and so one after another till they end with him at the guard-room door.

The Corporal of the new guard when he relieves the Sentries, is to examine whether the fentry-boxes, platforms, carriages of the cannon, palifades, &c. are in order or damaged. If be finds any of thofe things damaged, he is not to relieve that Sentry who had the charge of them, till the Town-Major is acquainted with it, otherwife he muft be anfwerable for the things fo damaged, or loft; but when he finds every thing in order, he is to relieve the Sentry. The Corporals are likewife to be attentive when the Sentries are delivering their orders to one another, left they fhould omit fome part of them.

As foon as the Sentries are relieved, the Corporal of the old guard returns with thofe relieved, forms them on the left of the guard, and acquaints his Officer with it. The Corporal of the new guard returns at the fame time, and acquaints his Officer with his proceedings, and whether he has relieved all the Sentries, or not, and if not, the reafons for it, that the Town-Major may be acquainted with it.

Where there are two Corporals on a guard, one of them is to take the charge of the guard-room, which he is to have delivered over to him clean, and the feveral utenfils belonging to it, in good order, by the Corporal of the old guard, which may be done while the others are relieving the Sentries; but where there is but one Corporal, it muft be done either before or after the Sentries are relieved; the Corporal muft take an exact account of what things are delivered to him, and the condition they are in, fince he muft anfwer for what are loft or fpoiled through carelefnefs.

It being a cuftom for the Drummers to take care of the Officers guard-room, with all the utenfils belonging to it, and to keep it clean, the Drummer of the old guard is to deliver it over to the Drummer of the new one in proper order.

When thefe things are done, the Officer of the old guard is to order his men to reft their firelocks, then
club;

club; after which, to close their ranks and files, wheel by divisions to the right, or left, according as they are to march from the place they are drawn up in, and then march off, the Drummer beating a *Troop*; for which reason, when a guard dismounts, it is called, *Trooping off of a guard*. When the men of the old guard club their firelocks, those of the new one are to be ordered by their Officer to rest, and his Drummer is to beat a *March*, when the other beats the *Troop*.

When the Officer of the old guard has marched his guard about an hundred yards, he may then dismiss the men, by first halting, and then dismiss them with the ruff of a Drum.

The Captain of the main-guard is to march to the grand-parade, and draw up his men before he dismisses them.

Those who command the guards in the outworks, are, when relieved, to march their guards an hundred paces within the gates before they dismiss them.

When a guard, which mounts with Colours, is relieved, or ordered to be dismissed, the men are not to club their firelocks, but march with shouldered arms, and the Drummers to beat a march to the grand-parade, and draw up; after which, the Officer who commands the guard is to send the Ensign to lodge the Colours with a proper Detachment to guard them, and to remain on the parade with the rest till he returns with the Detachment; after this, he orders them to club their firelocks, and then dismisses them; but if the place where the Colours are to be lodged is at a considerable distance, or that the weather is very bad, he may then order the Ensign to dismiss his detachment as soon as he has lodged the Colours; in which case he dismisses the rest, as soon as the Colours are marched out of sight.

As soon as the old guards are marched off, the new guards are to be drawn up on the ground where the old ones stood; after which, the Officers may order the men to ground their arms, or place them against

the

the wall of the guard-room, or lodge them in it, according as the conveniency of the place, or the weather will permit.

When a Sentry has not been relieved by the Corporal, for the reasons already mentioned, that of suffering any thing to be lost, &c. the Officer of the guard is to send the Corporal to acquaint the Town-Major with it; on which, he is to send one of the Town-Adjutants to enquire into the damages done, order the Sentry to be relieved, and sent to the Provost Martial's, or place appointed for prisoners, till he can be tried in a regular manner for the same.

When the guard-rooms are damaged, or the utensils belonging to them lost, or that the gates of the town, and barriers, or the draw-bridges, are spoiled, or out of order, the Officer is not to relieve the guard, till he has sent to the Town-Major to acquaint him with it, otherwise he will be obliged to repair those damages.

ARTICLE VI.

The manner of relieving Sentries, with directions how they are to behave on their posts.

When the Corporal goes to relieve the Sentries, as soon as he comes within six paces of the Sentry who is to be relieved, he orders his men to halt, and then rest their firelocks; the Sentry, who is to be relieved, is to rest his arms at the same time; the Corporal then orders the first who is to go on duty, to relieve the Sentry; upon which he recovers his arms, and advances within a pace of the Sentry, then halts, and rests his arms, and receives the orders, relating to that post, from the Sentry who is to be relieved; to which the Corporals are to give attention, lest some part of the orders might be omitted. As soon as they have delivered their orders, they both recover their arms, and exchange places, then rest their arms again. This being done, the Corporal orders the other men to shoulder, at which the two Sentries do the same, and the Sentry, who is relieved, falls in the rear of those who are going to relieve; then the Corporal marches to the next post.

If the Sentries are required to have their bayonets fixed, when the orders are delivered, they are to poise their firelocks, rest on their arms, and he who relieves, draws his bayonet, and fixes it, and he that is relieved, unfixes his bayonet, and returns it; then they recover their arms, exchange places, and perform all the other motions above-mentioned.

All Sentries are to be vigilant on their posts; neither are they to sing, smoke tobacco, nor suffer any noise to be made near them. They are not to sit down, lay their arms out of their hands, or sleep; but to keep moving about their posts, if the weather will allow of it.

They are to have a watchful eye over the things committed to their charge, and not suffer any of them to be removed, or taken away, till they have orders from the Corporal of the guard for it.

They are not to suffer any one to touch or handle their arms; or, in the night-time, to come within the reach of them.

They are not to suffer any light to remain, or any fire to be made near their posts in the night-time. Neither is any Sentry to be relieved, or removed from his post, but by the Corporal of the guard.

No-body is to strike or abuse a Sentry on his post; but when he has committed a crime, he is to be relieved, and then punished, according to the rules and articles of war.

When a Sentinel is taken ill on his post, or that the cold is so great that he cannot support himself under it, he is to call the Corporal of the guard, and acquaint him with it, in order to his being relieved; but when the Sentry, so taken, is at too great a distance to be heard by the Corporal, the next Sentry to him is to pass the word for the Corporal of the guard, and so from one to another, till it comes to the guard; for which reason, it is a standing rule to post Sentries within the call of one another, particularly on the ramparts of a town.

When Sentries have orders to stop people in the night-time, in order to their being examined, or to make

Chap. XI. *Military Discipline.*

rounds stand, as soon as they come within twenty paces of them, they are to challenge boldly, *Who comes there?* If the persons so challenged do not answer, but approach, the Sentries are to make ready their arms, and challenge a second time; and if they still advance without answering, they are to cock their firelocks, and challenge a third time; and if they advance after that without answering, the Sentries are then to fire, and return to their guard, if they find it necessary; otherwise they may continue at their posts, and load again immediately; but when the persons challenged, answer, the Sentries are to order them to stand, and call the Serjeant of the guard.

A Sentry on his post in the night, is to know no-body but by the counter-sign. When he challenges, and is answered *Relief*, he is to order them to stand, by saying, *Stand Relief, advance Corporal!* upon which the Corporal halts the men, and advances alone within a yard of the Sentry's firelock (first ordering his men to rest their firelocks, on which the Sentry does the same, as is directed in the relieving of Sentries) and gives him the counter-sign, taking care that no other person shall hear it; after which the Relief goes on in the manner before mentioned.

All Sentries, except those at the guard-room door, when they challenge, and are answered, *Round*, or *Patrole*, they are to say, *Pass Round*, or *Patrole*, and to rest their firelocks till they are passed, and not suffer them to come within the reach of their arms.

When two Sentries are placed at one post, which is always done at advanced posts, they are to be very attentive, and keep a profound silence: and when they hear any noise, such as the march of horse or foot, or any number of men approaching towards them, one of them is to return immediately to the guard, and acquaint the Officer with it, but without any noise, and then go back to his post. The Sentry who remained, is to listen with great attention to the noise, in order to discover what it was, and to make his firelock ready, and stand upon his guard, that he may not be taken by surprise; and when any person or persons come near him, and will not answer or stand when he has challenged and commanded them to do it, according to the foregoing directions, he is to fire, and return to his guard.

The Sentry at the guard-room door is to challenge briskly when any person comes within twenty paces of him; and if he is answered, *Round*, he is then to say, *Stand Round!* and rest his firelock, and call the Serjeant of the guard. He is not to suffer the Round to approach after that, till ordered by his Officer.

When an Officer goes to visit the Sentries, the Sentry is to challenge when he comes within ten or twelve pace, of his post; and when he is answer, *the Visit*, he is to say, *Stand Visit, advance one with the counter-sigd!* upon which he rests his arms, and permits the person, who is to give the counter-sign, to approach within a pace of the muzzle of his firelock, that none else may discover the counter-sign; therefore all counter-signs for foot ought to be a name, word, or number, and that to be spoken very near, and no louder than is necessary for the Sentry to hear; it being easy in the night for the enemy to approach undiscovered, near enough to distinguish a hem, whistle, or slap on the pouch, should they be given for counter-signs, and, by that means, be able to impose on the Sentry, and seize him, and after that surprise the guard.

A Sentry who is found sleeping on his post, or attempts to deliver it up to the enemy, or suffers it to be surprised through negligence, is to be punished with death; therefore all Sentries must be very alert, that they may avoid falling into these enormous crimes, since the articles of war, and the constant practice of all nations, make it absolutely death to those who shall be found guilty of them.

When the counter-sign is changed during the night, the Sentries are to take it from none but the Corporal of the guard.

When a town is besieged, or that they are apprehensive of the enemy's making some attempt to surprise them, it is customary for the Sentries posted on the ramparts

Chap. XI. *Military Discipline.*

parts to call out, every half hour, with a loud voice, *All is well*; when this is ordered, the Town-Major is to assign the post it shall begin at, and which way it shall go round. Upon the first saying, *All is well*, the next to him is to say the same, and so from one to another, till it comes quite round to him who began it. The design of this is to keep the Sentries alert on their posts, and to prevent their falling asleep. The Sentries at the guard-room doors are to be very attentive to the word (*All is well*) coming round; and when they find that it does not come punctually to the time, they are to acquaint their Officers with it, who are to send a Corporal with a file of men round their Sentries, lest any of them should have fallen asleep, or quitted their posts, in order to find out where it stopped, that the offender may be brought to punishment.

The word going round in this manner, is never used but in time of danger; or now and then to instruct young or unexperienced troops in their duty: for when things of this nature, which should be only practised on proper occasions, are constantly used when there is no necessity for them, they grow so familiar, till at last they fall into contempt, and, perhaps, neglected, when there is a real occasion for their being punctually observed. I am therefore of opinion, that it should not be used but for the reasons above-mentioned, that of real danger, and to instruct young Soldiers.

I believe what I have already said, relating to the duty of a Sentry on his post, will be sufficient to give any one a full and clear idea of it: however, I do not pretend to say that this is all, since particular cases will require particular orders: and without they are mentioned, it is impossible to give the necessary directions; but whatever orders a Sentry shall receive, whether those above-mentioned, or others, he is to execute them with the utmost exactness, since the safety of an army, or the preservation of a town, may often depend on the due performance of his orders.

CHAP. XII.

Consisting of instructions to the Officers on guard, from the time of mounting till they are relieved; with the manner of going and receiving rounds, and sending patroles; with the design of them.

ARTICLE I.

NO Officer is to leave his guard during the time he is on duty, which in garrison never exceeds twenty-four hours, but must send for what he wants.

He must not suffer above two men at a time to leave the guard, and then only for their victuals and drink; when they return, he may allow two more to go off on the same account; they should be allowed no more time than what is absolutely necessary, that each may have his turn; which, if they transgress, the Officer should punish them for it at their return. But lest some of the men should ask leave just before they are to go to Sentry, in order to escape or avoid their duty, the Officer of the guard should always send for the Corporal before he gives a man leave, that he may inform him when he is to go Sentry; as also to order the Serjeant or Corporal to set down their names, with the hour they went, and the time allowed them: when they return, they are to acquaint their Officer with it, that he may know whether they have been punctual or not.

The Officers of the port-guards are to examine all strangers who come into the garrison, taking their names in writing, with the place where they are to lodge, and the time they intend to stay: which they are to mention in the next report they send to the Captain of the main-guard,

guard; but when a person of distinction comes into the town, the Officer of the port-guard is to send an account of it in writing immediately, by an orderly man, to the Captain of the main-guard, who is to acquaint the Governor, or Commandant, with it as soon as he can. When any person comes into the town whom they have reason to suspect, by his not being able to give a good account of himself, the Officer is to send him to the Captain of the main-guard, who is to secure him, till he can acquaint the Governor with it, in order to his being further examined.

ARTICLE II.

The Officers of the port-guards are to send a report night and morning, in writing, to the Captain of the main-guard, in which they are to insert the names of all strangers who have come into the town, the place where they lodge, and the time they intend to remain, and those who go out of the town; as also of every thing that shall happen on their guard: which reports are to be signed by the Officers, specifying the day of the month, and the port it came from, and to be sent by the Serjeants who go for the keys to shut and open the gates.

All the other guards, except the reserve, are to send their reports in the same manner, and, at the same time, to the Captain of the main-guard. These are all called the ordinary reports, as being sent constantly, night and morning, at a fixed time.

Those which are called extraordinary reports, are only sent when any thing extraordinary happens on or near a guard, or a person of distinction comes into town, that the Captain of the main-guard may acquaint the Governor with it immediately.

As soon as the Captain of the main-guard receives the night-reports, he is to write them over fair in a sheet of paper, or more, if requisite, putting the report of each guard distinctly by itself, with the Officer's name who commands it: after which he is to sign it; and when the gates are shut, and the orders are given out, he is to wait on the Governor, give him the parole, and deliver him the report of the whole.

The Captain of the main-guard is to enter the morning reports in the same manner, with every thing that has occurred during the night, either relating to the several rounds or patroles, with the time each went and finished, that it may be known whether the Officers have complied with their orders, or not; as also what prisoners are on the main-guard, with the reasons for their being committed; and whether Soldiers, townsmen, or strangers taken up on suspicion. In short, he is to put every thing down which has happened between the evening report, and the time of relief, in order to give a faithful and exact report to the Governor, which he is to do as soon as he is relieved, by giving him the parole first in his ear, and then deliver him the report.

When any thing happens on any of the guards between the morning report, and the time of relief, such as strangers coming into town, &c. the Officers are to send an account of it to the Captain of the main-guard, that it may be entered with the rest, before he delivers it to the Governor.

When any of the rounds neglect going, or do not perform it at the hour appointed, the Officers of those guards to which the round or rounds have not gone, or gone after the time directed, are to mention it in their morning report to the Captain of the main-guard, who is to enter it in that which he gives to the Governor, that the reason for such neglect may be enquired into.

The reserve-guard being only a number of men kept in a readiness, to act either in the town, or to march out of it, as the Governor shall have an occasion for their service; the Officer who commands it, is therefore to receive no orders but from the Governor, or the Town-Major, by his directions, which he is to be ready to execute at a minute's warning. He is therefore to keep no more Sentries than what are necessary for the security of his guard, and only to patrole near his own guard-room;

neither

neither is he under the direction of the Captain of the main-guard, nor to make any report to him; but when he is relieved, he is to wait on the Governor, give him the parole, and deliver him a report of his guard in writing, signed.

ARTICLE III.

The Officers of the port-guards are to keep the barriers shut, and the draw-bridges up, on Sundays and Holidays, during the time of divine service; as also every day from twelve o' clock till one.

They are likewise to shut the barriers, and draw up the draw-bridges, at the approach of any party of armed men, though it should be Detachments of their own garrison, and acquaint the Captain of the main-guard with it immediately, that he may wait on the Governor to receive his orders for their admittance, without which they must not be permitted to come into the town. One Officer, or a Serjeant, may be allowed entrance, to shew the order or route, that the Governor may have an exact account of them.

When any Detachment, or a number of armed men, enter the town, the Officer of the port-guard is to have his men under arms; and if it is a Detachment commanded by an Officer, the men of the port-guard are to rest their arms, and the Drummer to beat a *March*; provided the party which enters beats a *March*; but if it is only a Serjeant's party, the guard is to remain shouldered, and the Officer remains at the head of it without his Esponton in his hand. This may be looked upon, by some, as too great a compliment from an Officer's guard to a Serjeant's party; but they must know that it is not done by way of respect to those who enter, but for the security of the town; lest the enemy, by having forged or procured a route or order, might send such a party to seize the gate, while the body lay concealed at some little distance, and ready to advance on the first signal. It is therefore a standing rule in all garrisons,

for

for the port-guards to be under arms, when any number of of armed men march into the town, though they belong to the garrifon.

When a fire breaks out in a garrifon, the Officers of the port-guards are to put their men immediately under arms, and order the barriers to be fhut, and the draw-bridges drawn up, and to keep them fo till the fire is extinguifhed.

This precaution is abfolutely neceffary in frontier garrifons, otherwife towns might be eafily furprifed, fhould the gates be left open on fuch an occafion ; it being natural for every body to run to that part which is fet on fire ; which might be contrived on purpofe by the enemy, by procuring proper emiffaries to do it, and who, by lodging troops, at the time appointed, within a proper diftance of the town, might, during the confternation, which always attends fuch accidents, feize one of the gates, and by that means poffefs themfelves of the town ; but by the fhutting the barriers, and raifing the draw-bridges, that danger will be effectually prevented, and leave them no room for fuch an undertaking, at leaft with any hopes of fuccefs.

When a riot, or a tumultuous affembly, happens near a port, the Officer of that guard is to ufe the fame precautions, in fhutting of the barrier, drawing up the bridges, and keeping his men under arms till it is over, for the reafons above-mentioned : but when thefe things happen to be only fome fmall diforder, occafioned by a quarrel, he may fend a Serjeant and file of men to quell it.

When a riot happens in thofe parts of the town which are at a diftance from the ports, the Captain of the main-guard is to fend parties, both from his own and the horfe-guard, to difperfe the mob, and feize the offenders.

In all frontier garrifons, it is neceffary to double the guards on market-days, and to examine ftrictly all covered waggons, or thofe loaded with hay or ftraw ; as alfo boats, barges or fhips, and every thing in which

men,

men, arms, or ammunition, may lye concealed; and when any thing of that nature is discovered, they are to stop it, and acquaint the Captain of the main-guard, that he may inform the Governor of it, and receive his directions.

ARTICLE IV.

Half an hour before the gates are to be shut, which is generally at the setting of the sun, a Serjeant and four men must be sent from each port to the main-guard for the keys; at which time, the Drummers of the port-guards are to go upon the ramparts, and beat a *Retreat*, to give notice to those without, that the gates are going to be shut, that they may come in before they are. As soon as the Drummers have finished the *Retreat*, which they should not do in less than a quarter of an hour, the Officers must order the barriers and gates to be shut, leaving only the wickets open; after which, no Soldier should be suffered to go out of the town, though port-liberty should be allowed them in the day-time.

The Town-Major, or, in his absence, one of the Town-Adjutants, must take a Serjeant and twelve men from the main-guard, and go to the Governor for the keys of the town, bring them from thence to the main-guard, and deliver them to the Serjeants of the several ports, who are to carry them to their guards, escorted by the men they brought with them. As soon as the Sentinels at the ports perceive the Serjeants coming with the keys, they are to give notice of it, on which the Officers are to turn out their guards, ranging the men under the vault or arch of the port, in two ranks, facing one another, that the keys may pass between them. As soon as the Serjeants arrive with the keys, the Officers are to order their men to rest their firelocks, and the Drummers to beat a *March*, till the gates are locked. He must order a Corporal and four men more with arms to escort the keys to the outermost barrier, and to place two men with rested arms, on every draw-bridge, till they

they return from locking the barriers. He must send likewise a sufficient number of men without arms to assist in the locking of the gates and drawing up the bridges.

When there are any guards to be posted in the outworks during the night, the Town-Major, or one of the Town-Adjutants, should go along with the keys of that port from whence they are to be detached, in order to see them posted, and to give the Officer or Serjeant who commands them, the word, counter-sign, and the necessary orders relating to the care of the post or posts to be guarded, and then see the gates of that port immediately locked.

When there are guards to be placed in the outworks at different parts of the town, and that the Town-Major and his Aids cannot see them all posted themselves, without keeping the gates open beyond the usual time, the Town-Major may send directions to the Officers of the port-guards, from whence they are to be detached, to go and post them, with the orders, parole, and counter-sign in writing, sealed up, to leave with those who command them, with directions not to open it till the gates are shut. As cases of this nature seldom happen, I do not know that the above method was ever practised; and therefore will not recommend it, but when it cannot be avoided by the night-posts in the outworks being too numerous for the proper Officers to see them all posted themselves; but whenever this should be the case, I believe the expedient will not be thought improper.

When the gates are shut, which the Officers on the port-guards are always to see done, the keys are to be carried back to the main-guard, by the Serjeants and Escorts who brought them, and delivered to the Town-Major, or Adjutant, who, when they are all returned, is to carry them to the Governor's, escorted by a Serjeant and twelve men from the main-guard.

As soon as the gates are shut, all the additional Night-Sentries within the walls are to be posted, and to take possession of all other night-posts which shall be ordered; after which the Officers are to order their men to recover
their

their arms, and lodge them in the guard-room, taking care to place them in such order, that every man may take his own firelock, when commanded, without any bustle or confusion.

The Serjeants who carried the keys back to the main-guard, are to remain there till they have received the night orders from the Town-Major, and the tickets for the rounds from the Captain of the main-guard: after which they are to return to their guards, and deliver the orders, parole, and counter-sign, with the tickets, to their Officers, and then to the Corporals of the guards.

As soon as the gates are shut, and the keys returned to the Governor, the Town-Major should come to the main-guard, and deliver out the night-orders to the Majors and Adjutants of the garrison, and to the Serjeants from the port-guards, and others.

The Captain of the main-guard is to deliver to the Serjeants from the port-guard, as many tickets as there are rounds ordered to go, taking care that the names of the Officers guards are named on the tickets, one of which is to be delivered to every round as they pass.

In frontier garrisons, they commonly order so many rounds as to have an Officer always walking on the ramparts in the night. When this is necessary, they compute the time that the first round will be going round the town; and when that has almost finished, the second is to begin, and so one after another, till the *Reveille* beats. These are called the Visiting-rounds. The Officers who dismount in the morning, are always appointed to go these rounds, because they are farthest from duty. They are to assemble at the main-guard at the time of delivering the night orders, to draw by lot for the hour each is to go his round at; after which, the Town-Major is to enter their names, Regiments they belong to, and the time of going their rounds, in his book; that if the Governor should find by the morning's report, that no round went such an hour, or staid beyond the usual time, he may inform him who should have gone then, that the reason may be enquired into.

The

The *Tat-too* is generally beat at ten o'clock at night in the summer, and at eight in winter. It is performed by the Drum-Major, and all the Drummers of that Regiment, which gives a Captain to the main-guard that day.

They are to begin at the main-guard, beat round the Grand-parade, and return back, and finish where they began. They are to be escorted by a Serjeant and a file of men from the main-guard.

They are to be answered by the Drummers of all the other guards: as also by four Drummers of each Regiment in their respective quarters, if the town is very large.

The *Tat-too* is the signal given for the Soldiers to retire to their chambers, to put out their fire and candle, and go to bed. The publick houses are, at the same time, to shut their doors, and sell no more liquor that night.

In frontier garrisons, the burghers are constantly obliged, when they go out, after *Tat-too*, to carry a light with them. Those who do not, are taken up by the patroles, and kept prisoners all night upon the guard, in order to be punished next morning by the Governor, for disobeying the orders of the garrison.

ARTICLE V.

The patroles are to go every hour in the night, from the beating of the *Tat-too* till the *Reveille*. The patroles are commonly composed of a Serjeant and six or twelve men from each guard. They are to walk in the streets to prevent disorders, or any number of people assembling together, and to oblige all those who keep public houses to send away their guests, and shut their doors. When they see any light in their Soldiers caserns or barracks, to oblige them to put it out, or acquaint the guard of those quarters with it, that they may see it done. To take up all the Soldiers they find out of their quarters: as also all the inhabitants who go without lights, if the orders of the garrison are such, and carry them prisoners to the guard. When any of the public houses entertain company after the patrole has forbid them, they are to carry the landlords to the guard, that the Governor may punish them the next day for their disobedience.

The Town-Major is to assign a proper district for each guard to patrole in, by dividing of the town in such a manner, that every street may be included in one patrole or another. The districts should lie contiguous to the several guards, that the patroles may not interfere with one another. The middle of the town belongs to the main-guard, and the streets near the ramparts to the port-guards.

It is the custom, in some garrisons, for the Horse-guard to perform these patroles on horse-back. When the town is very large, it will be very proper to order them to patrole through the principal streets of the town, and the great squares and market-places, to prevent any tumultuous assembly, or rising of the inhabitants; but as to the performing of the other parts, for which patroles are designed, as above-mentioned, how is it possible for them to comply with it? For as the noise of the horses feet will be heard at a considerable distance, it will be easy for those who disobey the orders of the garrison to avoid the patrole, and thereby escape due punishment: for which reason patroles of horse, in towns, are generally laid aside, except in the case above-mentioned, and those of foot appointed in their room; which, as being more useful, are infinitely more proper.

When the patroles are challenged by the Sentries, they are to answer *Patrole*; upon which the Sentry replies, *Pass patrole!*

When they return from patroling, and are challenged by the Sentry at the guard-room door, they are to answer, *Patrole of the guard*, naming it, as main-guard, reserve, or such a port; upon which the Sentry permits them to go into the guard-room and lodge their arms.

As soon as the patrole returns, the Serjeant is to make a report to his Officer of every thing that happened during his patrole, and what prisoners he has brought to the guard, that he may examine them himself, and set down their names in writing, the time and reason for their being taken up, the place of abode, if towns-men, or if Soldiers, the Regiment and Company they belong to; all which must be inserted in the morning report to the Captain of the main-guard, at which time the prisoners must be conducted there also.

ARTICLE IV.

The ordinary rounds are three. The Town-Major's round, the grand round, and the visiting round.

The extraordinary rounds, are those which are appointed to go every hour of the night, or every two hours, as the Governor shall think proper; which rounds are performed by the Officers who dismount the guard that morning, and are called the visiting rounds, as before-mentioned.

As soon as the gates are shut, and the night orders delivered to the garrison, the Town-Major may begin his round; the design of which is, that he may see whether all the gates are shut, the additional night-posts and Sentinels posted, and the Officers and Soldiers all on their guards.

Manner of going the rounds, and receiving them.

When the Town-Major goes his round, he comes to the main-guard and demands a Serjeant and four or six men to escort him to the next guard; and, when it is dark, one of the men is to carry a light. He may go to which gate first he pleases; whereas, all the other rounds, except the Governor's or Commandant's, are to go according to the method prescribed them.

As soon as the Sentinel at the guard-room door perceives the round coming, he should give notice to the guard, that they may be ready to turn out when ordered; and when the round comes within twenty paces

Chap. XII. *Military Discipline.* 209

of the guard, he is to challenge briskly; and when he is answered by the Serjeant who attends the round, *Town-Major's Round*, he is to say, *Stand Round!* and rest his firelock; after which, he is to call out immediately, *Serjeant, turn out the guard, Town-Major's Round.* No round is to advance after the Sentinel has challenged and ordered them to stand.

Upon the Sentinel's calling, the Serjeant is to turn out the guard immediately, drawing up the men in good order with shouldered arms, and the Officer is to place himself at the head of it, with his arms in his hand. After this, he is to order the Serjeant, and four or six men to advance towards the round, and challenge. When the Serjeant of the guard comes within six paces of the Serjeant who escorted the round, he is to halt and challenge briskly: The Serjeant of the escort is to answer, *Town-Major's Round*; upon which the Serjeant of the guard replies: *Stand Round; advance Serjeant with the parole!* and then orders his men to rest their firelocks. The Serjeant of the escort advances alone, and gives the Serjeant of the guard the parole in his ear, that none else may hear it, and while he is giving it, the Serjeant of the guard holds the spear of his halberd at the other's breast. He then orders the Serjeant to return to his escort, and, leaving the men he brought with him to keep the round from advancing, goes to his Officer, and gives him the parole he received from the Serjeant; the Officer finding the parole right, orders his Serjeant to return to his men, and then says, *Advance Town-Major's Round!* and orders the guard to rest their arms; upon which the Serjeant of the guard orders his men to wheel back from the center and form a lane, through which the Town-Major is to pass, the escort remaining where they were, and go up to the Officer, and give him the parole, laying his mouth to his ear. The Officer holds the spear of his Esponton at the Town-Major's breast while he gives him the parole. The reason of this ceremony, is, I presume, lest he should prove an impostor, and come to betray the guard; and that if he should

P   give

give a wrong word, or appear not to be the person whose character he assumes, the Officer may be prepared to punish him as he deserves; as also to be in a state of defence, lest he should attempt his life; the surprise of which might throw the guard into such confusion, for want of an Officer to command it, that the men would, perhaps, abandon their post, or deliver themselves up, without making any resistance, on the appearance of the enemy's troops, or a body of armed men advancing towards them: so that unless the above precautions were taken, *viz.* That of obliging the round to stand at some distance till the guard is put under arms, with all the other parts of the ceremony, as above-mentioned, it might be easy for the enemy to surprise an out-post or camp-guard, by lodging a party of their men at some convenient place near them, and then send out a small party in the nature of a round, with an enterprizing person to command it, and assume the character, in order to kill the Officer; which, with the enemies appearing that moment, would effectually prevail, and make them yield without any considerable resistance, or abandon their post; but the strict examination they are to go through before they are suffered to approach the guard, makes the enterprize too difficult to be attempted with safety to the person who shall undertake it. Besides, it is a standing rule, both in camp and garrison, for an Officer on guard to know no-body in the night, but by the parole; and till that is given in the usual form, he is to suffer none to approach his guard.

Though an enterprize of this nature would be more difficult in a garrison than in camp, it might, however, be effected, were these ceremonies laid aside: for, if the enemy can draw the inhabitants into their interest, they may send men into the town on market-days in the disguise of peasants, on pretence to sell provisions (it being impossible for the port-guards to distinguish who are really peasants, and who are not) with directions how they are to conduct themselves till the time appointed for the undertaking. We may suppose that the inhabi-

tants,

tants, who have entered into the plot, will not be wanting on their part, to make the necessary preparations of arms and ammunition, and to lodge them in some house near the gate, which they propose to seize; and to conceal the men who are sent in, till the time it is to be executed. When the time appointed comes, the enemy will send a body of troops superior to those in garrison, and take care to conceal themselves in the day-time, and not approach the town till night, and even not then till the appointed hour, for fear of being discovered; with full directions how they are to proceed, and when to advance to the gate which is to be seized.

Were measures of this kind well concerted, both within and without the town, it would be no difficult matter to seize a port-guard, and, with proper instruments, break open the gates, let down the draw-bridges, and give a free entrance to the enemy, before the troops of the garrison could be got together to prevent it.

This digression may be thought foreign to the present subject, and therefore might have been omitted; but my design in it, is to shew young gentlemen the necessity there is, for the ceremony in going and receiving rounds; sending frequent patroles; Sentinels not suffering any one to come within the reach of their arms; none permitted to come near a guard at night, till they are strictly examined; the searching of waggons, boats, &c. which come into the town, lest men, arms, or ammunition, should be concealed; strangers who enter, giving an account of themselves, and obliging the inhabitants to give an account of all strangers who lodge in their houses, without which no frontier town could be safe from the enemy, in conjunction with the inhabitants, who may always be gained by the force of money, or at least a sufficient number to carry on the design; so that the danger within is to be guarded against, as well as that without; and how is it possible to be done, unless the foregoing rules and ceremonies are strictly adhered to, and duly executed?

The Town-Major having given the Officer of the guard the parole, he is then to examine if the gates are locked and well secured; whether they have taken possession of their night posts, and placed the additional night Sentinels, and count the men who are under arms, to see if they are all on guard, and if not, to enquire into the reason of their absence. He may likewise enquire into the night orders, as also all others relating to the guard, that if there should be any mistake in them, he may then rectify them. After these things are done, he should send back the Serjeant and men, who attended him, to the main-guard, and take the same number from this guard to escort him to the next; and so from one guard to another, till he has finished his round. He is to be received at all guards in the same manner as he was at the first.

As the Town-Major's round is designed to see if the gates are locked, the night posts posted, and the orders delivered right, I presume, he may go either along the ramparts, or through the streets, from one guard to another, as he shall think proper; but all the other rounds, except the Governor's, must go along the ramparts.

As soon as the round is gone, the Officer is to order his men to lodge their arms; and when the Serjeant returns from conducting any of the rounds, he is to acquaint his Officer with it, and whether the Sentinels, as they passed, were alert or not.

When it was said, that the Town-Major is to go his round when the night orders were delivered, it is to be understood, that he is not to go till they are, and that the gates are shut; after which, he may take his own time, there being no certain hour prescribed him, provided he goes and finishes before twelve o'clock. Besides, it is even necessary for him to go at uncertain hours, and change his way of going, in order to keep the guards alert: however, he must always go the first round, to verify the night orders.

When the Town-Major has finished his round, he is to wait on the Governor, give him the word, and make him a report of the state of all the posts, and the condition he found them in.

In

Chap. XII. *Military Discipline.*

In the *French* garrisons, all the Officers who command guards are to give the parole to the Town-Major, or, in his absence, to the Town-Adjutant when he goes the first round, which is always called the Town Major's round, though gone by one of his Aids. The reason for this is not by way of compliment to the Town-Major; but, by receiving the parole from the Officers of the guards, he may know if they have received it right, otherwise, they say, how can he be certain if they know it? When the Town-Major goes any more rounds that night, he must give the parole to the Officers every time he goes, except the first, as the other rounds do.

This method, in my opinion, is grounded on a very just principle, and therefore preferable to the other, which is that of the Town-Major's giving the word to the Officers on guard, even the first time of his going: however, I am not going to introduce new customs here; but only set down the practice abroad; and, where they differ from the *Dutch*, from whom we have taken the greatest part of our discipline, by having been in a long alliance with them; and though it must be owned, that we could not have followed a more perfect system of discipline than theirs, both in camp and garrison; yet, in particular parts, though perhaps not many, one may be allowed to dissent from them, and prefer those of a neighbouring nation, when we find them better; as I think that is, just above-mentioned, of the Officers giving the Town-Major the word, in his first round, in order to verify it.

All other rounds must be received in the same manner as is directed for the Town-Major's; only with this difference, that the Officers on guard are to give the parole to the grand round; but all other rounds are to give it to them: and though the Governor should go his round after the grand round is made by the Captain of the main-guard, he is to give the word to the Officers on guard: but, in this case, the Governor may carry an Officer to give the word for him.

The Captain of the main-guard is to go the grand round, which is commonly made about midnight; and the Lieutenant is to go the visiting round, which is made about an hour before day.

When the Governor intends to go the grand round, he is to send notice of it to the Captain of the main-guard, to prevent his going it, and that he may be prepared to receive him; it being usual for the Governor to come to the main-guard first, and take an escort along with him from thence to the next guard, or to conduct him quite round if he thinks proper, and order the Lieutenant of the guard to attend him. The Governor may order what number of men for his escort he pleases, which generally consists of a Serjeant and twelve men.

When the Governor goes the grand round, the Captain of the main-guard is to go the visiting round.

The grand round, or any round which the Governor, or Commandant, shall make, may begin where they please, because whatever round they meet, is to give them the word; whereas, when two other rounds meet, that which challenges first has a right to demand the word of the other. But as this might occasion disputes in the giving the word, should both challenge together, or imagine they did, the place where they are to begin, and the hour which each round is to go at, must be particularly mentioned; by which method they cannot possibly meet, but will follow one another in a regular manner, provided they are punctual in the execution.

All rounds (the Town-Major's, Grand, and Governor's rounds excepted) are to demand a ticket from the Officer of each guard, as they pass it; and when they have finished their rounds, they are to deliver them to the Captain of the main-guard, who is to examine them very carefully, to see if they have missed any of the guards; after which, he is to set down the Officers names that went the rounds, and the hours they returned at; as also every thing that happened extraordinary to them in the going their rounds; such as Officers being absent

Chap. XII. *Military Discipline.*

absent from their guards, or negligent in their duty; Sentinels drunk, asleep, or off their posts; if they discovered any thing from the ramparts, or heard any noise in the country; or saw any number of people assembled together in the town, or found any disturbance, that he may mention it in his report next morning to the Governor.

When a round discovers from the ramparts any number of troops, or hears any considerable noise which may induce them to believe there are some near, they must give notice of it to the Captain of the main-guard, who is to acquaint the Governor with it immediately, that he may send the reserve to strengthen that post from whence the noise was heard, and to give orders for the troops to repair immediately to their Alarum-posts, to prevent the garrison's being surprised.

If the Sentries on the ramparts make the same discovery, they are to call the Corporal of the guard, and acquaint him with it, who is to inform his Officer, that he may enquire into it immediately; and, if he finds it of any consequence, he must send an account of it to the main-guard, that the Governor may be acquainted with it; after which, he is to draw out his guard, that he may be ready to oppose any attempt that shall be made, either from without, or within the town. He should send a Corporal and two men round his Sentries, to see that they are all alert, and to give them strict orders to be very watchful; and when they discover any thing further, to pass the word for the Corporal, that he may come and know what they have discovered: the Corporal must have orders to go on to the next port-guard, and acquaint the Officer with what they saw or heard; who is immediately to put his guard under arms, send his Corporal round his Sentries, and acquaint the next port-guard with it, that they may do the same; and so from one to another, till it has gone quite round.

This precaution, of sending to every guard, is absolutely necessary, since it is certain, that when the enemy have a design to surprise any place, they will endeavour to draw your attention from it, by making a show of attacking some remote part, and thereby draw your forces from the real attack; therefore prudence directs us to be careful of the whole, and not suffer ourselves to be amused by appearances; but to suspect a deeper design, than what may at first offer itself to our view.

The Officers of the guards should likewise send out patroles, lest the inhabitants should be in concert with the enemy: which they will easily discover by their assembling together, and by that means avoid the danger from within, or be prepared against it.

The design of rounds is not only to visit the guards, and keep the Sentries alert, but likewise to discover what passes in the outworks, and beyond them; for which reason, the Officers who go the rounds, should walk on the banquet, and go into the Sentry-boxes, that they may look into the ditch, and discover with more ease what passes there; they must likewise enquire of the Sentries if they have discovered any thing on their posts, or heard any noise.

Some Governors have a round go just before the opening of the gates; their reason for it is this: as it is then pretty light, that round, by the elevation of the ramparts, will be able to discover a good way into the country. When they have no horse in garrison to patrole, it is not only proper, but absolutely necessary. When such a round is ordered, it should begin just at *Reveille*, which is then so light, that they may see an hundred and fifty yards, and when the officer has finished his round, and returned to the main-guard, the keys are to be sent to the gates.

ARTICLE VII.

In frontier garrisons, as soon as the *Reveille* beats, the Officer of the horse-guard, with his men mounted, is to repair to the main-guard, and, according to his orders from the Captain, he must go, or send patroles of horse

Chap. XII.   *Military Discipline.*   217

horse out of the gates to reconnoitre the country. The patroles are generally composed of a Corporal, and four or six Troopers each.

At the beating of the *Reveille*, a Serjeant and twelve men from the main guard are to attend the Town-Major, or one of the Town-Adjutants, to bring the keys from the Governor's to the main-guard, and to deliver them to the Serjeants from the ports, who are to be there, with four men each, at the same time; and after they have received them, they are to carry them to their guards, followed by their patroles of horse, who are to be let out to reconnoitre.

Upon the Serjeant's coming with the keys, the Officers are to have their guards under arms in the same manner as is directed for the shutting of the gates When the keys are come, the Officer orders the first gate to be opened, and lets out the patrole of horse, and then shuts it; when that is done, the draw-bridges are let down, and as soon as the patrole has passed them, they are drawn up again; after that, the barriers are opened, the horse patrole let out, and closed again after them, the Corporal and four men of the guard remaining within the barrier.

The time which the horse are to patrole, and the parts which they are to reconnoitre, must depend on the orders they shall receive from the Governor; however, they generally reconnoitre those places from whence they apprehend the danger may proceed; such as the roads leading to the garrisons or territories of the enemy; hollow ways, woods or thickets, which lie near the garrison, lest troops should be concealed there; and to get upon the eminences which overlook the country, in order to discover what passes a considerable way in it; and when they have complied with their orders (which takes them up generally three quarters of an hour, unless their situation requires them to be more circumspect) they return to the garrison, and acquaint the Officers of the port-guards with what they have discovered; and if all is well, the Officers then order the gates to be opened, the

draw-

draw-bridges to be let down, and the patroles to enter, who return to their own guard, and make a report to their Officer; and when they are all returned, he is to make his to the Captain of the main-guard, which report he is to enter with the rest. As soon as the gates are opened, the keys are to be carried back to the main-guard, in the usual form, and delivered to the Town-Major, and from thence escorted to the Governor's.

During the time that the horse are patroling, and that the keys remain at the ports, the guards are to continue under arms; the Corporal and four men who opened the barrier to let the horse out, are to shut the barrier after them, and to remain within it till they return. As soon as the draw-bridges are drawn up, after the horse have passed them, the wicket of the innermost gate should be opened, that the men who are placed at the draw-bridges may come and give the Officers an account when the patroles return; upon which the Officers of the guard must go and examine the Corporals of the patroles themselves; after which, if they have no reason to the contrary, they are then to order the gates to be opened, the draw-bridges to be let down, and the patroles to enter; but till the Officers have examined them, the gates are not to be opened, or they admitted.

CHAP.

CHAP. XIII.

Of Detachments, visiting the Soldiers quarters, and the Hospital.

ARTICLE I.

ALL Detachments which are sent from a garrison, are either to guard posts which lie at a distance from the place, for escorts, or parties.

The Detachments are composed of an equal number of Officers and Soldiers from the Regiments in garrison, as is done for the town-guard; but is a separate duty from it, and a roll is kept apart by the Town-Major.

When a Captain, Lieutenant, and Ensign, are commanded, they order the eldest Regiment to give a Captain, the second Regiment the Lieutenant, and the third Regiment the Ensign, with an equal proportion of men from each Regiment in the garrison. When another Captain is ordered, he is taken from the second Regiment, and so on till every Regiment has given one, and then it begins again with the eldest. The same method is observed by the Lieutenants, Ensigns, and Serjeants.

There is no roll kept for the Drummers, because the Regiments send as many Drummers as they have Officers ordered on duty: so that whatever guard, or detachment, an Officer is posted to, he takes his own Drummer along with him; which is a standing rule, both in camp and garrison.

Every Battalion, whether strong or weak, furnishes an equal proportion of Officers and Soldiers to all Detachments; whereas, in the town-guards, they are so

far

far indulged sometimes, as to give only in proportion to their numbers, when the disproportion between them is very considerable; occasioned by the loss of men in action, violent sickness, or from the different establishments as to Numbers: for it would be highly unreasonable that a Battalion of 400 men should do equal duty with one of 600. But if their establishment is the same, and that they have not one of the above reasons to plead, or that the difference in numbers is but inconsiderable, they must then do the town-duty equal with one another, as well as all Detachments.

The ordinary complement of a Captain, when detached, is a Lieutenant, Ensign, two Serjeants, two Drummers, and forty-five or fifty men, Corporals included.

A Lieutenant has a Serjeant, Drummer, and twenty-five or thirty men.

An Ensign has a Serjeant, Drummer, and twenty or twenty-four men.

A Serjeant has twelve, fifteen, or eighteen men; but on Detachment always eighteen men.

When a Serjeant is detached from a garrison, he should never have less than eighteen men; because it is a rule with the *French*, and their neighbours, never to send out a party of a smaller number; and whenever they seize a party under it, they treat them like Party-Blews, or robbers, unless they should have lost some of their men, which will appear by their order or route.

A Major's command is from one hundred and fifty men to three hundred.

A Lieutenant-Colonel's from four hundred men to six hundred.

A Colonel's from six hundred men to eight hundred, or a thousand.

But notwithstanding the ordinary complement above-mentioned, yet an Officer must not scruple to march with a much smaller number, when the King's service requires it; they are likewise often commanded with a greater number, than what is mentioned: however,

when

when the command comes up to, or exceeds a thousand men, they appoint General Officers to command them, in proportion to the number detached, all which must depend on the will of the Governor, or Commandant of the place.

They generally send a Lieutenant-Colonel and a Major along with a Colonel when he is detached; but always one Field-Officer with him, if there are any; when a Colonel is commanded, he takes his own Adjutant with him.

ARTICLE II.

If an Officer, after he has marched his Detachment beyond the outermost barrier of the place, should be ordered to return, it passes for a duty; but if he should be ordered back before he has passed the barrier, it will not be allowed as a duty; but must go with the next that is commanded; for his marching off from the parade does not excuse him, as a great many imagine: it holds the same with the private men.

If an Officer's tour of duty to mount the Guard should come while he is on Detachment, he is not obliged to take it when he returns; but if he comes into town the day that he is to mount, and that his Detachment is dismissed before the guards are marched off from the parade, he must then mount the guard, without any difficulty or scruple.

All Parties or Detachments of infantry which are sent out of a garrison, should not consist of less than nineteen men, that is a Serjeant and eighteen private Soldiers: but, unless there is a necessity for their sending no more, they should not detach less than an Officer and twenty-five men.

All parties are to have a passport, or order, in writing, signed by the Governor or Commandant of the garrison, and sealed with his coat of arms; in failure of which, should they be taken, they will not be treated like prisoners of war, but left to the discretion of the enemy;

enemy; and the Governor to whom they belonged, has no right to claim them.

If a party under nineteen men are taken, and have not a pass under the Governor's hand and seal to shew that there were so many detached, they will be treated as above-mentioned, and condemned by a court-martial, either to the gallies for life, or a punishment equally as bad.

This custom, I presume, is only to prevent a smaller number from being detached, who can only be sent to pillage and steal, which is looked upon, by all sides, as an ungenerous way of making war, since it can only make a few people unhappy, without contributing any thing to the service, or the bringing the war to a conclusion.

It is likewise to prevent Party-Blews, which are parties of robbers, who sometimes dress themselves like Soldiers, and plunder every one they meet, without distinction; for which reason they are always hanged by both sides when taken: therefore, to distinguish real parties from those, it is absolutely necessary that they should have passports signed and sealed by the Governor, or Commandant of the garrison.

Another reason, why smaller parties than nineteen men ought not to be sent, may proceed from the danger of their being overpowered by a Party-Blew, or the peasants.

Though the rules abroad declare positively, that any party which shall be taken, consisting of less than nineteen men, shall be treated like Party-Blews; yet, I think, it can never be taken in the literal sense, but must mean, that if they are sent out with less, they will be treated in that manner: for as action, sickness, or desertion, may reduce a greater party under that number, it would be the height of barbarity to use those ill, who remained, for a crime which they were not guilty of: therefore it cannot be doubted, in my opinion, but the producing their passport will clear them, and make them be used like prisoners of war.

When

Chap. XIII. *Military Dicifpline.* 223

When an Officer who commands a party, is obliged to send a Detachment from his party, either out of his sight or call, he is not to send less than a Serjeant and eighteen men; to whom he must give an exact copy of his passport, with directions what they are to perform, and the time they are to return to him, or the garrison, writ under the said passport. It is therefore presumed, that an Officer who has not the command of forty men or upwards, will not send a party out of his sight or call, otherwise he will be left with a smaller number than the custom of war allows of.

All parties or detachments must return to their garrison punctually at the time appointed; unless they are prevented by the enemy's getting between them and home, and, in order to avoid them, are obliged to retire to the next place of safety, or go a great way about for fear of being taken: in which case, an Officer is not only excusable for staying beyond the time, but deserves thanks for his care and conduct.

The time limited, with the number of Officers and Soldiers, is always inserted in the pass or order; and for the most part, the service which they are employed on, is likewise mentioned: however, that may be committed to a particular paper, and only communicated to the commanding Officer of the Detachment, when the service they are sent on requires secrecy.

ARTICLE III.

There must be a Serjeant and a Corporal of each Company orderly for a week, the Serjeants and Corporals taking it in their turns to perform this duty; which, however, does not excuse them from mounting of the guard, or going on party, when their tour comes for either. When this happens, the Serjeant or Corporal, who is to be on the orderly duty next, must perform the orderly duty for them, till they are relieved, or that their orderly time expires; after which they commence the orderly duty for themselves: neither are the orderly

Ser-

Serjeants or Corporals, who are on guard or party during their orderly week, obliged to repay the orderly duty, which is done for them during that time, when they return; it being a fixed rule for those who are next in turn to perform it, without being repaid it.

The orderly Serjeants and Corporals are to receive all orders which shall be delivered, either to the garrison in general, or to the Regiment and Company they belong to in particular, and deliver them to their Officers.

They are to march the men of their Companies, who are to mount the guard, to their Captain's quarters, to be viewed by one of their Officers, before they are sent to the Regimental-parade; and if they carry any man that has not his arms, ammunition, clothes, and accoutrements in good order, they are answerable to their Officer for it, and liable to be punished for the same; in order to avoid which, they must inspect these things before they present the men to their Officers. They are to march them from thence to the Regimental-parade; and deliver them to the Adjutant of the Regiment, who is to examine the men again, to see that they are in order, and that he has his complement. After this, they are to attend the Adjutant till the guards are marched from the Grand-parade, that if any orders are to be delivered that morning, either from the Governor, or their Colonel, they may be ready to receive them.

They are to remain constantly at their caserns or barracks, when they are not employed in receiving or delivering of orders, and carrying their men to the parade, as abovementioned, that they may be ready to execute all commands which shall come at any time from the Governor or their Colonel.

They must take care to keep six or eight men of a Company, of those who are to go first on duty, always in the way, in case there are any ordered for parties, or any other occasion, for which they may be wanted; and when any men are ordered, they must see that their arms, ammunition, &c. are in good order, before they go to the parade, and take care to provide them with

am-

Chap. XIII. *Military Discipline.*

ammunition-bread and pay for the number of days they are to be out on party, or that proportion of both which shall be ordered for them. The Corporals are to march the men to the parade, and deliver them to the Adjutant, and the Serjeants may remain at their barracks, (unless they are employed in getting the money and bread for them) there being no occasion for their going with any men to the parade, but these who mount in the morning, without the rules of the garrison order it otherwise.

They are likewise to see that the men keep the caserns or barracks very clean and in good order, and that the utensils belonging to them are neither spoiled nor lost. They are to make the men sweep their rooms very clean every morning, and make their beds; and afterwards to wash themselves very clean, and dress in a Soldier-like way, by having their shoes well blacked, their stockings and cravats well rolled, their hats cocked, their hair dressed, and their clothes brushed and put on to the best advantage; but till these things are done, they are not to suffer them to leave their quarters, that they may not appear slovenly in the streets.

They are to call over the roll of their Companies as often as it shall be ordered, and make a report of the absent men to their own Officers and the Adjutant, that they may be punished for it.

They must go through every room immediately after *Tat-too*, and oblige the men to put out their fire and candle, and go to bed.

The men of each Company should be divided into messes, each mess consisting of four or six men, or according to the number in each room: and every pay-day, each man should be obliged to appropriate such a part of his pay to buy provisions, which money should be lodged in the hands of one of them, in order to be laid out to the best advantage, which the orderly Serjeants and Corporals are to see duly executed, and make each mess boil the pot every day. Without this is carefully looked into, the Soldiers will be apt to spend their pay on liquor, which will not only occasion their neglect of duty, but, in all probability, the loss of a great many men by sickness, for want of proper victuals to support them. It is therefore a duty incumbent on every Officer, to be more than ordinarily careful in this particular, and not to think themselves above the looking into these things, since the preservation of their men depends so much on it : for in those Regiments where this method is duly observed, the men are generally healthful ; but when it is neglected, great numbers fall sick and die.

The Captains should visit their mens quarters at least once a week, and the Subalterns twice, to see that they are kept clean and in good order ; as also to inspect into the several messes of their Companies, and to see whether their provisions are good, and the money laid justly out.

In some Regiments there is an Officer appointed daily to visit the caserns or barracks of the Regiment, to see that they are kept clean, and that the men dress their victuals, and to make a report of the whole to the Colonel ; however, that should not prevent the other Officers from looking into it also.

The Major should visit the whole very often, that he may know whether the other Officers do their duty, and reprimand those who neglect it ; it being his immediate business and duty, to see all orders punctually obeyed.

ARTICLE IV.

In all garrisons, there is an Officer of a Regiment ordered to visit their sick men daily. They are to examine nicely into the manner their men are treated; and if they are kept clean ; what medicines and diet they have given them, taste their bread and broth, and see their proportion of bread and meat, and enquire of them how they are used ; a report of which, with the number belonging to each Company, they are to make to their Colonels.

Besides

Chap. XIII. *Military Discipline.*

Besides this Regimental inspection, the Governor appoints an Officer or two of the garrison to visit the hospital, to see how the men are treated, with the number belonging to each Regiment, and to make him a report of the whole.

A Captain, who has any regard for his men, will not think these general visits sufficient; but will go from time to time himself, and enquire into their state, and send his Officers and Serjeants to do the same. There are a great many little things which may save the life of a poor sick Soldier, and which they cannot have but from their own Officers; so that unless they go to see them, they may perish for the want of them; therefore common humanity requires this duty of us even to strangers, but much more so to those who are immediately under our care, and share the danger of the war with us.

There is one rule which should be strictly observed, which is, not to suffer the men to continue too long in the hospital; but to take them from thence as soon as possibly they can: for it often happens, when they stay too long after they are recovered, that they relapse and die; occasioned by the badness of the air, which must be, in some degree, infected by the breath of the sick, in spite of all the care that can be taken to prevent it. But the least evil that can happen by their remaining too long in the hospital is, that they will thereby contract a slothful, lazy, idle habit, and turn, according to the military phrase, *Malingerors*; that is, men who have lost all spirit to the service, and feign themselves sick when there is a prospect of action, or that they are to undergo any hardship or fatigue, in order to be sent to the hospital; which life, through habit, becomes agreeable to them.

When there is no publick hospital, there should be some rooms in the quarters appointed for the sick men, to which they may be removed, lest they infect the others; and that the Surgeon of the Regiment may attend them with the more eafe. They fhould likewife have nurfes appointed to attend them, and proper care taken about their diet: and, unlefs the diftemper is of a very malignant nature, I am of opinion, that they fhould not be fent to the hofpital upon every flight indifpofition, but removed into the infirmary of the Regiment, the air of which muft be much purer than the other, and therefore there is a greater probability of their quick recovery. Befides, the evil above complained of will be avoided; which is an article of no fmall confideration to the fervice.

CHAP.

CHAP. XIV.

Relating to the command of the Governor in his own town, with the respect and obedience due to him from the Troops which compose the Garrison; as also what compliments are to be paid to all the other Officers.

ARTICLE I.

WHOEVER is Governor of a town, has the entire command of the troops which compose the garrison, though Officers of a superior rank to him in the army should be ordered in with them: for the town being committed to his charge, he is answerable to his master for it, and consequently cannot give up the command without express orders from him in due form, or from him to whom he shall delegate his power.

In the absence of the Governor, the command devolves on the Lieutenant-Governor: and if the Town-Major has a commission of Town-Major Commandant, (which is sometimes conferred on those abroad) the command falls to him in the absence of the Governor and Lieutenant-Governor; otherwise it goes to the eldest Officer in the garrison, whether he is of the horse, foot, or dragoons, who is called, during the time, Commandant of the garrison. This is the general rule; but as they may be obliged, on particular occasions, to throw a considerable body of troops into the garrison (either for the defence of it, or to annoy the enemy) and that a General Officer of a considerable rank may be ordered in with them, it is usual to give him a commission of Commandant of the troops, in the body of which is particularly

particularly specified, how far his power over them is to extend, to avoid all disputes that might happen betwixt him and the Governor about it: and though this may in a great measure, lessen and divide the Governor's power, yet the outward marks of distinction are generally left with him, such as the giving the parole, the administration of the civil affairs, keeping the keys of the town, &c. as also the signing of the capitulation, jointly with the Commandant of the troops, in case of a surrender.

The reason for appointing a Commandant of the troops, I suppose, may arise from the Governor's not being of a rank in the army sufficient to give him a due authority over them, or that he may not be thought equal to the command; but, supposing him equal to it, both from his experience and ability, unless he is distinguished with titles of dignity, his orders will not be so readily executed as if he was: and though a commission of Governor, creates him, in a manner, Captain-General in his own town, yet when Officers of an equal rank to him in the army are ordered into the garrison, it is a hard matter for him to keep up his command as it ought to be, or get them to obey him with the same deference as they would one of a superior rank; and, if it proves so, when only those of an equal rank are commanded into the garrison, it would be much more difficult for him to exert his authority over those who are his superiors in the army, as well as shocking to them to be commanded by an inferior; the truth of which, with the detriment that arises from it to the service, is so well known in *France*, that when the case happens so there, and that they have no mind to supersede the governor, they always appoint an Officer of rank and ability (in proportion to the number of men, which, upon occasion of danger, shall be ordered into the garrison) Commandant of the troops; in which case care is generally taken, that the person so appointed be of such a rank in the army, that not only all dispute about command in relation to him is out of the question, but likewise any contests of this kind that

may

Chap. XIV. *Military Discipline.* 231

may arise in the garrison are terminated, and his decisions more readily submitted to, than if they came from one of an inferior character. I shall now proceed to the command of a Governor, when there is no Commandant of the troops appointed.

How far the Governor's power extends over the civil, must be determined by the laws and constitution of the country: however, all persons in the town, whether ecclesiastical or civil, are subject to his jurisdiction, as far as it relates to the order and preservation of the town; and whoever offends therein, though he may not have the power of punishing, yet he may secure their persons till they can be tried in a regular manner for the crimes they have committed.

His power over the military is very extensive; for all the Officers and Soldiers in the garrison are obliged to obey him without controul.

He may order the troops under arms as often as he shall think proper, either to review them, or upon any other account.

He may send out detachments, or parties, without being obliged to give a reason to the Officers for it, or come to an explanation with them on that head. Neither have they a power to demand it; but if they think themselves grieved, they may represent it to him in a respectful manner; that is, singly, and by way of request, and not in a riotous way, and in numbers, since that will be deemed mutiny, which, by the articles of war, is death.

Neither Officer nor Soldier must lie a night out of the garrison without the Governor's leave; but that the Colonels, or those who command Regiments, may have a proper authority over their own corps, a Governor seldom grants his leave of absence to either Officer or Soldier, but at their request. A Governor, who has a true notion of the service, will act according to this rule; and it appears to me reasonable that he should do so, otherwise, how can they answer for their Regiments, if their Officers and Soldiers have leave of absence given

them

them without their knowledge? Besides, as the Colonels are supposed to have a thorough knowledge of those under their command, they must be proper judges who ought or ought not to have leave given them, and therefore will not importune the Governor but when it is reasonable they should have it; which will not only ease him of a great deal of trouble, but likewise prevent his being imposed upon, by their pretending to have business, when perhaps pleasure, or the love of idleness, is the chief motive which induces them to ask it; the truth of which cannot be so easily entered into by the Governor, as the Colonels, who, in justice to their Regiments, will limit the number they ask leave for, that the duty may not fall too hard on those who remain.

What is above-mentioned, without entering into the deference due to Colonels, when it relates to those immediately under their command, is so equitable, that it is generally followed: but however just this rule may appear, yet a Governor has an undoubted right to deviate from it when he shall think proper, by granting his leave of absence to either Officer or Soldier, without the consent of their Colonels: and though particular Regiments may suffer now and then by such a proceeding, yet that evil is of less consequence to the service, than what the limiting of the Governor's power might produce, *viz.* the loss of subordination: which is of such weight and consideration, that it is the very life and soul (if I may be allowed the expression) of discipline, without a due Observance of which, the service can never be carried on; for, whoever endeavours to weaken it, by making the Officers or Soldiers independant of the principal persons who are placed over them, whether Governors or Generals, must do it either through an evil design or ignorance, since both produce the same effect, disorder and confusion; a state which Soldiers may be easily brought into, (from a natural love of independency which reigns in all mankind) but not so soon remedied; for, when a licentious, independent humour has prevailed amongst troops, it must be time, infinite pains, and severity, to reduce them to their proper obedience; the want of which may prove as prejudicial to the state, as the want of troops; since the loss of subordination produces not only the neglect of orders, but, in a great measure, the power, or at least an imaginary one, to dispute them; the consequence of which is too well known to be farther enlarged upon.

The practice of the army in this case is, that when an Officer has business that may require his absence from the garrison, he is to make his first application to his Colonel, and to desire him to intercede with the Governor for leave; and if the Colonel complies with the Officer's request, he should wait upon the Governor in his behalf; but if the Colonel refuses the Officer, he may then, no doubt, apply to the Governor; though such a step should not be taken without he is necessitated so to do, either from extraordinary business, or that he finds himself hardly used by his Colonel, since the doing it is, in a manner, putting him at defiance, and therefore not to be rashly undertaken.

When any of the private men want leave, they are to apply to their Captains first, the Captains to the Colonel, and, if he agrees to it, he is to send their names by the Adjutant to the Town-Major, that he may acquaint the Governor that they have his consent, and to desire he will be pleased to grant them his leave of absence.

When the Soldiers have applied to their Captains, and are refused by them, they may then apply to their Colonels; but they ought not to do it till they have been with their Captains, for the same reason that an Officer ought not to apply to the Governor till he has been with his Colonel.

ARTICLE II.

All Soldiers who have leave to go out of the garrison, must have passports signed by the Governor, specifying the Regiment to which they belong, the place they are to go to, and the time they have leave to be absent; the particulars of which must be given in by the Adjutant to the

the Town-Major. Whoever goes without one of these passports, or is found taking a contrary road to that which is expressed in it, will be looked upon as a deserter, and, when taken, tried accordingly. It is therefore the duty of the Officers on the port-guards, to examine all Soldiers who shall come into the town, and do not belong to the garrison; and when they find any of them without a pass, or that they have taken a wrong route, or have any reason to suspect it forged, they are to send them to the main-guard, in order to their being further examined by the Governor, or those whom he shall appoint for that purpose; and if they are found to be deserters, they should be secured, till they can be sent to their Regiments to be tried as such.

When Officers on party meet any Soldiers, they must examine their passports: and if they have any reason to suspect them, they must take them prisoners, and deliver them over to the main-guard when they return to their garrison, and acquaint the Governor with it.

No Regiment can hold a Court-martial, or punish any of their men, without first obtaining the Governor's leave, or the Commandant's in his absence: however, it is customary, upon the first application which the Colonel makes of this kind to the Governor, to give him a discretionary power to hold Regimental Courts-martial, as often as he shall have occasion, and to put the sentence in execution, provided the Regiment is not to be under arms at the performing it; because no Colonel can order his Regiment under arms, either for exercise, punishing offenders, or otherwise, without having leave every time from the Governor: therefore, it is usual to punish the Soldiers on the Regimental parade, in the presence of the men who mount the guard in the morning, unless the sentence directs otherwise.

When the Colonel, or Commanding Officer, would have the Regiment under arms for exercise, review, or to punish any of his men, he must send the Adjutant to the Town-Major, that he may acquaint the Governor with it when he goes to receive the night orders; and

if

it granted, the Town-Major is to give out in publick orders, that such a Regiment is to be under arms, &c. to-morrow morning.

The ceremony of giving out in publick orders, when Regiments are to be under arms, has an appearance, as if it was only to keep up the authority of the Governor, and to shew his command over the troops in his garrison; and, indeed, I never heard any reason given for it, but that it was the custom: however, it cannot be doubted, but that a better reason than custom can be given for it; but since it has not come to my knowledge, I beg leave to offer my own opinion on that head.

Should a part of the garrison draw out in the morning without the rest being apprised of it, they might imagine that it proceeded from some attempt of the enemy, who were going to surprise the town, and consequently occasion their beating to arms: therefore, to prevent these false alarums, which would not only fatigue the troops, but, by their being too often repeated, make them dilatory in reparing to their Alarum-posts upon a real occasion; as also cause a bustle and disturbance in the town: it is therefore necessary, that it should be given out in orders by the Town-Major the night before, when any of the troops are to be under arms, that all may know it. Besides, the assembling of troops, without the Governor's leave, must put the town in the power of those Officers who command them; especially if we will suppose any ill intention, or correspondence with the enemy: for, though it is to be presumed, that Officers of their rank are above temptation, yet instances of the contrary may be given; and in war particularly, we ought not to rely on what they will not do, but on what they cannot do.

ARTICLE III.

An case of an alarum, the Officers and Soldiers, who are not on guard, are to repair, with their arms, immediately to their Alarum-posts.

Upon these occasions, the Colonel's Company may be ordered to assemble where the Colours are lodged, which is generally at the Colonel's quarters, to guard them from thence to the Alarum-post, of the Regiment.

Sometimes all the Field Officers Companies are ordered to assemble there; but unless the garrison is very numerous, they will be of more service with the Regiment, one Company being sufficient to guard them; the Ensigns who are to carry the Colours are to assemble there at the same time.

The reason for the troops being ordered to their Alarum-posts, may proceed from one of the three following causes:

First, Upon the appearance of the enemy before the town, or intelligence being brought, that a body of their troops are marching towards it; therefore to prevent a surprise, it will be proper to order the Regiments to repair to their Alarum-posts.

Secondly, Upon any considerable rising of the inhabitants, or tumult in the town, that the Governor may be able to disperse the mob, and bring the offenders to justice.

Thirdly, Upon a fire breaking out in the town, it is extremely necessary to have the troops at their Alarum-posts; for by their being assembled, they may be sent under the command of their Officers to assist in the extinguishing of it, and to keep the streets open that the engines may be brought to play; as also to keep the mob from stealing the goods which may be saved from the flames. Besides, as the town may be set on fire by a stratagem of the enemy, and, by lodging a body of troops at some distance from the town, they may endeavour to seize one of the gates, during the consternation, which, by the assistance of the inhabitants, might be easily effected, were the precaution of shutting the gates and assembling the troops omitted.

But on whatever occasion the alarum may be given, when the troops are assembled, no Colonel must dismiss his

his Regiment, though it should prove a false alarum, till he receives the Governor's or Commandant's orders for it. Thus far I have endeavoured to shew the command which a Governor of a town has over the troops in it, and how the Officers and Soldiers are to conduct themselves towards him on that head; and in the following article I shall shew the respect which is to be paid to him, and the other Officers who shall come into the garrison, by the guards.

ARTICLE IV.

All Governors whose commissions in the army are under the degree of General Officers, shall have, in their own garrisons, all the guards turn out with rested arms, and beat one *Ruffle*; and though the main-guard turns out with rested arms every time he passes, yet they give him the compliment of the Drum but once a day; but all the other guards beat as often as he appears near them.

If they are General Officers likewise, they are then to have the further compliments paid them, by the several beatings of the Drum, as is practised in the army, and are as follows.

To Generals of the horse and foot, the guards turn out, rest their arms, beat a *March*, and the Officers salute.

To Lieutenant-Generals, they turn out, rest their arms, beat three *Ruffles*, and the Officers salute.

To Major-Generals, turn out, rest their arms, and beat two *Ruffles*, but not salute.

To Brigadier-Generals they turn out with rested arms only; but of late they have added one *Ruffle* to the compliment.

To Colonels, their own quarter-guards turn out, and rest their arms once a day; after which, they only turn out with ordered arms.

To Lieutenant-Colonels, their own quarter-guards turn out with shouldered arms once a day, at other times they only turn out, and stand by their arms.

To Majors, their own guards turn out with ordered arms once a day, at all other times they stand by their arms.

When a Lieutenant-Colonel, or a Major, commands a Regiment, their own quarter-guards pay them the same compliment as is ordered for the Colonel.

All Sentries rest their arms to their own Colonel, Lieutenant-Colonel, and Major; but to those of another Regiment, they only stand shouldered, This ceremony is the same both in camp and garrison.

The main-guard is to rest their arms to the Governor, and pay him the compliment with the Drum, as before directed. If he continues to walk on the parade, or before the guard, they may lay down their arms.

All Sentries are to rest their arms as he passes them, or comes near their posts.

A General of the horse and foot, when in garrison, has a Serjeant and two Sentries at his door.

All Lieutenant-Generals have the same.

A Major-General is to have two Sentries at his door, and the same compliment paid him by the guards, as in camp.

A Brigadier is to have one Sentry at his door, and one *Ruffle* from all the guards in the garrison.

All Colonels, or Officers who command Battalions, are to have one Sentry, which they are to take from their own Regiments; but those Colonels who have no Regiments in the town, are to have the Sentry from the main-guard, or one of the port-guards, if their lodgings lie more convenient for them.

The main-guard is to turn out, and stand by their arms once a day to all Colonels; but all other guards must order their arms for them as often as they pass.

The main-guard is to pay no compliment to the Lieutenant-Colonel and Majors; but the other guards are to stand by their arms for them.

Lieu-

Lieutenant-Colonels are to be treated in their own garrifons as Colonels, and the Majors Commandant as Lieutenant-Colonels, unlefs their rank in the army entitles them to a greater compliment; but when either of them command the garrifon, they are then treated in all refpects as Governor.

When the Governor, Lieutenant-Governor, and the Major Commandant, are abfent, or by ficknefs rendered incapable of acting, the eldeft Officer in the garrifon is to take the command upon him, who is called Commandant of the garrifon, and has all the refpect paid him by the guards as Governor, except that of the Drum, unlefs his rank in the army entitled him to it before.

Thefe were the rules eftablifhed by King *William*; but of late the Governors who are General Officers have a *March* beat to them in their own garrifons: however, by the beft information I could get, I do not find that the Governors who are not General Officers, have exceeded the former compliment of one *Ruffle*.

A Captain-General of *Great-Britain*, a Marfhal of *France*, and a Velt-Marfhal of the Empire, being the chief military titles of thofe Kingdoms, which are given to fubjects, they are all treated with the fame degree of refpect, both in camp and garrifon.

When a Marfhal of *France* comes into any of their own garrifons, the ftreets are lined, by the troops, from the gate where he enters, to his lodging; the Soldiers reft their firelocks, the Drummers beat a *March*, and he is faluted by all the Officers and Colours. His guard, which confifts of a Captain, Lieutenant, and Enfign with Colours, and fifty men, are placed at his door before he comes into the town. He commands all Governors, and they are to receive the Parole from him. This is the Cuftom in *France*, and eftablifhed by the King's order.

How far this method may be confiftent in *England*, in regard to a Captain-General, is what I cannot determine, there being no regulation of this kind eftablifhed by the royal authority, that I know of; but as the late Duke of *Marlborough* (a copy of whose commission of Captain-General is hereunto annexed) had the same honours paid him in all the garrisons abroad, I presume he looked upon them as his due.

ANNE R.

*A*NNE, *by the grace of God,* &c. *To our right trusty, and right well-beloved cousin and counsellor,* John, *Earl of* Marlborough, *greeting.* WHEREAS *we have thought it necessary for our service to appoint and constitute a Captain-General for the commanding, regulating, and keeping in discipline our troops and land forces, which are, or shall be allowed by act of parliament to be raised and kept on foot:* KNOW YE, *therefore, that we, reposing especial trust and confidence in the approved wisdom, fidelity, valour, great experience and abilities of you, the said* John *Earl of* Marlborough, *have constituted and appointed, and by these presents do constitute and appoint you to be Captain-General of all our troops and land forces, already raised, and hereafter to be raised, as aforesaid, and employed in our service, within our kingdom of* England, *dominion of* Wales, *and town of* Berwick *upon* Tweed, *or which are, or shall be employed abroad in conjunction with the troops of our allies.* GIVING, *and by these presents granting unto you full power and authority, by yourself, commanders, captains, and other Officers, them to exercise, array, and put in readiness, and according to the provision of arms appointed for them, well and sufficiently cause to be weaponed and armed, and to take, or cause to be taken, the musters of them, or any of them, (by the commissary-general of the musters, or his deputies, or by such other officers as he shall assign for that purpose) as often as you shall see cause; and the said forces to divide into parties, regiments, troops, and companies, and with them, or any of them respectively, to resist all invasions which shall be made by our enemies, and to suppress all rebellions and insurrections which shall by levying war be made against us; and all enemies making such invasion, and rebels who*

who shall so levy war, and be found making resistance, to fight with, kill and destroy. As also with full power and authority for us, and in our name, as occasion shall require, according to your discretion, by proclamation or otherwise, to tender our royal mercy and pardon to all such enemies and rebels as shall submit themselves to us, and desire to be received into our grace and pardon. And we do likewise give and grant unto you full power and authority to hold, or cause to be held, from time to time, as often as there shall be occasion, according to your discretion, one, or more, military or martial court, or courts, in pursuance of, and according to the purport and true meaning of an act of parliament passed in the thirteenth year of the reign of our late dearest brother King William *the Third, of ever blessed memory, entitled,* An Act for punishing of Officers and Soldiers that shall mutiny or desert in *England* or *Ireland: And in the same court, or courts, to hear, examine, determine, and punish, all mutinies, disobediencies, departure from captains, commanders, and governors, according to the directions of the said act; and to cause the sentence or sentences of the said courts to be put in execution, or to suspend the same, as you shall see cause. To have, hold, exercise, and enjoy the said office of Captain-General, and to perform and execute the powers and authorities aforesaid, and all other matters and things which to your said office doth or may of right, belong, and appertain unto you, during our pleasure. Willing and commanding all officers, soldiers, and persons whatsoever, any way concerned, to be obedient and assisting to you our Captain-General, in all things touching the due execution of this our commission, according to the purport and intent thereof. In witness, &c. Witness, &c.*

Ex. Edw. Northey.

May it please your Most Excellent Majesty.

YOUR Majesty is hereby graciously pleased to constitute John Earl of Marlborough, Captain-General of all your land forces within your kingdom of England, dominion of Wales, and town of Berwick upon Tweed, or which are, or shall be employed abroad in conjunction with the troops of your allies, empowering him to execute all the powers and authorities thereunto belonging, during your Majesty's pleasure.

Signified to be your Majesty's pleasure by warrant under your royal sign manual, countersigned

By Mr. Secretary VERNON,

March 12, 1701-2.   EDW. NORTHEY.

ARTICLE V.

Whoever commands in a castle, fort, redoubt, or citadel belonging to the town, must send every day to the Governor or Commandant of the Town for the orders.

The same rules are to be observed by the garrison of the citadel as are given for those of the town; only with this difference, that the Governor of the citadel is not to suffer above one third of the Officers to be out at a time, though they should only desire to walk into the town. This is the method established in *France*. But as the citadels in that kingdom are built to be a check upon the towns, in order to keep the inhabitants in awe, this strict rule may be very proper, lest, by suffering the Officers and Soldiers to go out at pleasure, it might be surprised, when there remained only the ordinary guards to defend it. But where the case is not the same, I presume, the Governor may, in some degree, dispense with this order, by not adhering so strictly to it.

As there are separate Governors for the most part in *France*, the Governor of the town has no command over the Governor or garrison of the citadel: neither has he the liberty of going into the citadel without leave of the Governor of it: however, the Governor of the citadel is obliged to send every day to the Governor of the town for the parole, though his rank in the army should be superior to the other's. This may be thought absurd, that a superior must send to an inferior for the word; but thus it is established in *France*; and it is very common there, to find the Governor of the citadel an elder Officer, than the Governor of the town; which may proceed from the great dependance they have on their citadels, and therefore chuse Officers of considerable rank and experience for their Governors; and as there are no ill consequences attending it, by being no diminution to their rank in the army, they all submit to it.

ARTICLE · VI.

The Town-Major and the Town-Adjutants are to visit all the guard-rooms, caserns, and barracks pretty often, to see that they are kept in good order, and that the furniture and utensils belonging to them are neither lost nor damaged more than what may be reasonably expected. They are likewise to view all the parts of the fortifications, the Sentry-boxes, platforms, batteries of cannon, spare carriages, &c. and that the palisades are not stolen or decayed, and make a report of the same to the Governor, that those things, which are out of order, may be repaired in time.

In frontier garrisons, those who keep public houses must send an account in writing every night of all their lodgers to the Town-Major, specifying their names, quality, and country, when they come into the town, and from whence; that he may shew it to the Governor, in order to compare it with the night report from the Captain of the main-guard, by which he will know whether the Officers on the port-guards do their duty, in examining all ſtrangers who come into town, or the inn-keepers conceal any of their lodgers, or thoſe who came in gave a wrong account of the place where they were to lodge, by having ſome evil deſign to manage, that he may take proper meaſures for their being found out, and puniſhed according to their deſerts. In time of war, all private houſes are obliged to give an account to the Town-Major when any ſtranger lodges with them.

Where the towns are large, they have commiſſaries appointed to take an account of the ſtrangers from the public and private houſes, it being impoſſible for a Town-Major to perform this and all the other parts of his duty.

The following plan is a table for all the duty which is done by the Officers and Serjeants in a garriſon, and which is kept by the Town-Major, and the Adjutants of the ſeveral Regiments. As each Battalion gives an equal proportion of men, there is no occaſion for the inſerting of a column in the ſaid table for them.

A table for the ſeveral duties in a garriſon, to be kept by the Town-Major, and the Adjutants of the Regiments.

Regiments	Town-Guards.			Reſerve.			Detachments.			General Courts Martial.		Examination.		Viſiting of the Hoſpital	
	Captains	Subalt.	Serjeants	Captains	Subalt.	Serjeants	Captains	Subalt.	Serjeants	Captains	Subalt.	Captains	Subalt.	Captains	Subalt.

CHAP.

CHAP. XV.

Consisting of Camp Duty.

ARTICLE I.

BY camp duty, as I understand it, is meant, guards ordinary and extraordinary; though by taking it in its full extent, it will include every part of the service which is to be performed by the troops during the campaign; But as I have treated on several branches before, I shall now proceed to that which relates to the ordinary and extraordinary guards: and as a great part of the camp duty is performed in the same manner as that of a garrison, I shall only give directions for those parts in which they differ.

Guards ordinary, are such as are fixed during the campaign, and are relieved regularly at a certain hour every day; and are as follows. The grand guards of the cavalry, the standard and quarter-guards, the rear-guards, and the picquet guards of each Regiment; the guards for the General-Officers, train of artillery, bread-waggons, Quarter-Master-General, Majors of Brigade, Judge-Advocate, and Provost-Marshal.

Every Battalion of foot has for the quarter-guard, a Subaltern, two Serjeants, one Drummer, and thirty men; for the rear-guard, a Serjeant and ten or twelve men; and for the picquet-guard, a Captain, two Subalterns, three Serjeants, two Drummers, and fifty men. As every Battalion has the same constantly, the Majors of Brigade keep no detail of this duty. The complement of the other guards is as follows.

GUARDS.

	Captains	Lieutenants	Ensigns	Serjeants	Drummers	PrivateMen
The General in Chief has	1	1	1	2	2	50
General of horse and foot	1	1	1	2	2	50
Lieutenant-General of horse and foot	—	1	—	1	1	30
Major-General of horse and foot	—	—	1	1	1	20
Brigadier	—	—	—	1	—	12
Quarter-Master-General (as such only)	—	—	—	1	—	12
The Majors of Brigade, incamped together	—	—	—	1	—	12
Judge-Advocate	—	—	—	1	—	7
Provost-Marshal, as such, a Serjeant and eighteen men, but when he has prisoners, there is added a Subaltern, Serjeant, Drummer, and thirty men	—	1	—	2	1	48

The train of Artillery, according to the Number they shall require.

The guard which mounts on the General in Chief, has always Colours.

ARTICLE II.

Method of mounting and dismounting of the quarter and rear-guards.

As soon as the *Troop* has done beating, which is generally about nine in the morning, the men who mount the quarter-guard are to be formed into a rank entire, on the first or outermost line of parade, facing outwards, and directly in the front of the Colours.

The rear-guard likewise draws up in a rank entire, and in the rear of the quarter-guard.

When they are formed, the Adjutants are to deliver them to the Officers who mount, on which they are to place

Chap. XV.   *Military Discipline.*   247

place themselves at the head of their men, with their Espontons in their hands. After this, the Officer who commands the quarter-guard of the Regiment on the right of the line, orders his Drummer to beat a short *Preparative*, which is to be followed by the Drummers of all the quarter-guards which are to mount. This *Preparative* is to give them notice that they may be ready to march all at the same time. As soon as it is answered by all the Drummers on his left, he is to march his guard straight forward, on which the rest are to do the same, keeping an equal pace with him; and when they come within six paces of the old quarter-guards, they are to halt, then order them to rest their firelocks, after which the Officers advance towards one another, and the Officers of the old guards deliver their orders to those of the new. Then goes on the relief of the Sentries, delivering of the quarter-guard tents, &c. in the same manner as is directed in garrison-duty. While the Sentries are relieving, the Officers of the new guards are to face their men to the left, and open them to a proper distance, that the men of the old guards may pass between them.

The rear-guard marches off at the same time with the quarter-guard, through the interval of the Regiment, and is to observe the same directions, in relieving, as are given for the quarter-guard.

As soon as the Sentries are relieved, the Officer who dismounts on the right of the line, orders his Drummer to beat a *Preparative*, which is to be answered by all the other Drummers who are to dismount, after which they are to order the men of the old guards to club their firelocks and march, which they are all to do at the same time, taking their motions from that on the right of the line. When the old guards club, the new ones are to rest.

The Officers who dismount, are to *Troop* their guards to the first line of parade; and then halt; after which they are to order the men to rest their firelocks, recover their arms, and march and lodge them in their bells of arms. The Officers are then to make a report to the commanding Officer and Major of their own Regiments, of every thing that happened during the time they were on guard, with the names and crimes of the prisoners in writing, and by whom committed.

When the old quarter and rear-guards are marched off, the Officers of the new guards are to order their men to recover their arms, face to the right, and march into the ground where the others stood; after which to lay down their arms.

The same orders which were given to the guards in garrison about the Officers keeping their guards; their not allowing above two men to go off at a time; the relieving of Sentries, and how they are to behave themselves by day and by night; the receiving of rounds, and respect to be paid to the General Officers, must be punctually followed by the guards.

The quarter-guards are to be placed about seventy-four yards in the front, and directly opposite to the center of their own Regiments, facing them. The design of a quarter-guard, is rather for preserving the peace and tranquility within the Regiment, by quelling all disputes that may arise, either between Officer and Officer, or amongst the Soldiers, than for a security against the enemy: however, they are not to neglect that part neither, but to have a watchful eye to the front, left some of the enemies parties should pass the grand guards in the night, and fall upon them before they have time to prepare for their defence.

The rear-guard usually consists of no more than a Corporal and six private men in the day-time: but after the *Retreat* is beat, the Officer of the quarter-guard is to detach a Serjeant, and four or six men, as may be ordered, to remain with the rear-guard till morning.

The rear-guard must take care that no disorders are committed in the Sutlers tents or booths; must oblige them to put out their fire and candle in due time, and see that they entertain no body after the *Retreat*. They are likewise to take care that the horses belonging to the

Regiment are not stolen; and when any of them break loose, they are to stop them, and call those who have the care of them to catch them. When they find any Soldier or Soldiers drinking in a Sutler's tent at an improper time, they are to carry both the Soldiers and the Sutler prisoner to the quarter-guard, where they are to remain till the commanding Officer thinks proper to release them, or have them tried by a Regimental court-martial, in order to their being punished for their crimes.

An hour after the beating the *Retreat*, the Officer of the quarter-guard is to send a patrole of a Serjeant and six men round the Regiment, to see if the Sutlers have obeyed the above orders; to oblige the Soldiers to put out all their lights; and to visit all the Sentries, to see that they are alert on their posts; and if they find any one asleep, they are to secure his firelock, and send immediately to the Corporal of the guard to have him relieved, and committed a close prisoner, till he can be tried for his crime. When the Serjeant returns with the parole, he is to make a report of what happened, and what state he found every thing in, to his Officer.

These patroles should be sent every two hours, that is, an hour after each relief, during the night; so that with the relief and the patrole, the Sentries, and the several parts of the Regiment, will be visited every hour.

Upon any noise or disturbance in the Regiment, the Officer of the quarter-guard is to send a Serjeant, and a file of men, to enquire into the reason of it, and to put a stop to all disorders that may arise. If the Serjeant finds the Soldiers quarrelling, he must bring them prisoners to the guard; but if it is amongst the Officers, he must send immediately to his Officer, that he may come and confine them to their tents; and in the mean time the Serjeant must not suffer them to fight, which if they persist in doing, he is impowered to use force to prevent it, till the Officer of the guard comes. In short, the Officer of the quarter-guard is to have the same inspection over every thing that happens in the Regiment,

ment, as the Captain of the main-guard has over that of a garrifon.

ARTICLE III.

All the other guards ordinary, except the picquet, mount immediately after the beating of the *Troop*.

There are two parades for the forming of all guards or parties, that are done by detachments from each Regiment.

The firft is called the parade of the Brigade, and the fecond, the grand-parade.

The parade of the Brigade is generally at the head of the eldeft Regiment of each Brigade, and the grand-parade about the center of the firft line.

The Adjutants are to draw out the men, who fhall be ordered to mount, at the head of their own Regiments firft, and to examine into the condition of their arms, ammunition and accoutrements, and to fee that they are clean and well dreffed; after which they are to conduct them to the parade of the Brigade, and deliver them over to the Major of Brigade, and to wait there till he has looked into their ftate and numbers, that they may anfwer for what is wanting.

The Adjutants are to do the fame by all detachments that fhall be ordered from their Regiments, whether by day or by night, and not leave it to be done by the Serjeants-Major, as is but too frequently practifed.

The Majors of Brigade muft therefore be on the parade, to receive all detachments, that fhall be ordered from the Brigade, from the Adjutants of the feveral Regiments, and to oblige them to attend him till they are fent to the grand-parade.

Unlefs the Majors of Brigade are very punctual in performing this part of their duty, it is almoft certain that the Adjutants will be remifs in theirs, particularly in thofe which fhall be commanded out in the night: for when they know that the Major of Brigade will not be at the parade to form the detachment, they will be
apt

apt to lie in their beds, and order the Serjeant-Major to draw out the men, and march them to the parade; the confequence of which may prove detrimental to the fervice, both in the lofs of time, and for want of a due infpection into the men's arms and ammunition, fince we may naturally fuppofe, that the Corporals will not be quite fo diligent in drawing out their men for the Serjeant-Major, as for the Adjutant, nor take fo much care about their arms and ammunition, by not having the Adjutant prefent to look into it.

It is from fuch neglects as thefe, that a great many defigns mifcarry; for, let a fcheme be ever fo well concerted, one half hour's neglect, or lofs of time, in the executive part, may be fufficient to difappoint the whole, or occafion a much greater difficulty in the fuccefs. It is therefore abfolutely neceffary, that the Majors of Brigade fee all the detachments of their own Brigade paraded, at whatever time they fhall be ordered, and oblige the Adjutants to attend them till they are fent to the grand-parade.

When a Major of Brigade is of the day, he muft appoint one of the Adjutants of his Brigade, to fee all the detachments of it formed during the time he is on that duty.

The detachments are to draw up on the parade of the Brigade in the fame manner as the Regiments are encamped; thus: the eldeft on the right, the fecond on the left; and fo on from right to left, till the youngeft comes in the center.

As foon as the detachments from the feveral Regiments are paraded, and that the Major of Brigade has examined into their condition and numbers, he is to order the Officers, who mount with the men, to march them to the grand-parade: the particular method for the forming of them there, and detaching them from thence, fhall be mentioned in the following article.

ART I.

ARTICLE IV.

The Major of Brigade of the day is to be on the grand-parade, to receive the detachments from the several Brigades, in the same manner as each particular Major of Brigade do those from the several Regiments of their own Brigades; and to examine whether each Brigade has sent the number of Officers and Soldiers as was ordered.

The detachments from the several Brigades are not to be drawn up by seniority, but by lot, as directed for garrison duty: therefore, the Major of Brigade of the day must have as many lots ready as there are Brigades in the foot, and order a Serjeant of a Brigade to draw for them, according to which they are to draw up on the grand-parade.

As soon as the detachments are drawn up, the Serjeants are to draw up in the front of their own men, in the same manner as is directed in forming the guards in a garrison; after which, the Major of Brigade of the day is to tell off the several guards as the Town-Major does, by beginning at the right, appointing the Serjeants to them, and ordering the men, as they are told off, to order their arms. When all the guards are told off, the Officers are then to draw for their guards; but as this may occasion the Officers on the right of the line to mount with the men on the left, I am of opinion, that it would be more proper to place the Officers, as near as possible, to those guards to which the men of their own Brigades are detached: for as the Brigades are to draw every day for their posts on the grand-parade, it will hardly fall out, that the same guards will come to their share two days together; and therefore cannot fall harder on one Brigade than another. My reason why I think this method preferable to that of the Officers drawing for their guards, is, that when the guards are relieved, the Officers should march to the parade of the Brigade, and dismiss them there, instead of the grand-parade,

parade, by which means they will be difmiffed near their own encampment; whereas, by difmiffing them on the grand-parade, thofe men who are encamped on the extremities of the firft and fecond lines, will have a confiderable way to go to their Regiments, if the army is tolerably large, and thereby not only fatigue them, but throw the temptation of a Sutler's tent in their way, and, by getting drunk, lofe their arms and accoutrements, and the fear of being punifhed for the fame, may induce them to defert; but by the method I propofe, this inconveniency will be, in a great meafure, avoided, fince the men will be difmiffed near their own Brigade.

When the Officers are to draw for their guards, as is the general practice, they are, when relieved, to march their guards to the grand-parade, and difmifs them there.

The General Officer's guards, according to their feniority, are to be told off firft, then that for the Train, Provoft-Marfhal, &c.

As foon as the Officers are pofted to their guards, the men may be ordered to fhoulder their arms by beat of drum. After that, the Major of Brigade of the day is to order the guards to march off in the fame manner as is practifed by the Town-Major in a garrifon, and to fee them all march off from the parade, before he leaves it.

When an army is compofed of the troops of different Princes, thofe troops are commanded by General Officers of their own; in which cafe, the General Officers have guards from their own troops; fo that thofe guards do not come into the general detail of the army, but are kept a-part by the Majors of Brigade belonging to thofe troops; therefore thofe guards are not detached from the grand-parade, and confequently do not come under the cognizance of the Major of Brigade of the day. However, the above method will ferve for each nation, and their own Majors of Brigade muft take it day about to parade their own guards.

When it thus happens, the guards ordinary, which the Major of Brigade of the day is to detach from the

grand

grand-parade, are those of the train, Provost-General of the army, and the bread waggons; as also any other, for which the whole army is to give an equal proportion of Officers and Soldiers.

ARTICLE V.

The picquet-guard, as it is called, is a body of men who are to be always ready to march at a moment's warning, either to sustain out-posts, foraging escorts, or, in case the enemy should endeavour to surprise you in your camp, to march out and attack them, in order to give the army time to draw up.

The number which every Battalion gives to the picquet, is mentioned in the first article. When the picquet is ordered to march, another is immediately ordered to supply their room, in case a second should be commanded out.

The picquet-guard continues on duty only twenty-four hours, and is drawn out at the head of each Battalion every night, in the following manner:

The Officers and men for the picquet-guard, being ready dressed and accoutred, as soon as the *Preparative* for beating the *Retreat* is made, the men take their arms and form in the streets before their tents: there the orderly Serjeants and Corporals, having their arms likewise, examine them, and form those of their respective Companies into three ranks, within the lines of their tents. When the *Retreat* begins, they march them forward, the front rank even with the bells of arms, each orderly Serjeant, or Corporal, being three paces advanced before the men of his Company. The Officers, Serjeants, and Drummers, go to the head of the Colours, and taking their arms, wait there.

As soon as the *Retreat* is ended, the Adjutant orders,

Advance to form the picquet!

Upon this they march forward in three ranks to the lines

lines of parade, the Officers, Serjeants, and Drummers of the picquet, as well as the orderly Serjeants, or Corporals, advancing twelve paces before the line of parade; as soon as they are all at their ground, the Adjutant orders;

Halt!

Upon this the Officers, Serjeants, and Drummers, face to the right-about.

Form the Picquet!

At this word of command, the whole, excepting the Officers, Serjeants, and Drummers, face to the right and left inwards.

March!

They march together, closing to the center; and the Officers, Serjeants, and Drummers, take their posts. The orderly Serjeants and Corporals close likewise; but so as to be opposite to the men of their Companies, to answer for what may be wanting, or amiss.

Halt!

The picquet faces to the front, and the orderly Serjeants, or Corporals, to the picquet.

The Adjutant then goes through the ranks, and, after examining the whole, orders the orderly Serjeants, or Corporals to their Companies, to call the rolls: they are to march regularly thither, facing to the right and left outwards. He then is to acquaint the Captain, that his picquet is complete.

The Captain, together with his Officers, is then to examine the men's ammunition; and then orders;

Prime and load!

Which they are to do regularly and together. As soon as the Colonel, or Field-Officer of the picquet, has acquainted the Captain, that he may return his picquet, the Captain (having first cautioned them to be ready to turn out on the first notice) orders,

Picquet to the right and left to your Companies!

Upon which the Officers, Serjeants, and Drummers, move three paces to the front, and the men face to the right and left outwards.

March!

They march till they come opposite to their bells of arms, waiting for the next word of command,

Halt!

Upon this, they face to their bells of arms.

March, and lodge your arms!

They march together with an equal pace, and lodge their arms carefully, the Officers, Serjeants, and Drummers doing the same.

The eldest Serjeant of the picquet is to get a list of the mens names immediately, and give it to the Captain, that if any one is wanting, when the picquet is ordered out, they may know who it is, in order to his being punishing for neglect of duty; as also to prevent the men being changed, or ordered out upon any other command, while they are on the picquet; for should the men be changed, how should the Officers of the picquet know whether their mens arms were in order or not, or that they were provided with ammunition? for which reason,

they

Chap. XV.   *Military Discipline.*   257

they should always order those men on the picquet-guard who came last off duty, that others may not be commanded on guards or detachments out of their turns, by having those men on the picquet, who are the first on the roll to go on duty.

The quarter-guards are to turn out, and remain with shouldered arms during the time that the picquet continues at the head of the Regiments; and when the picquet is turned in, the men of the quarter-guard are to lodge their arms, either in their bell of arms, or shed erected to keep them dry.

Most Regiments have a distinct roll of the picquet duty for the Officers, from that of guards ordinary and extraordinary, which roll begins with the youngest, as the other does with the eldest, that the Officers may have an equal share of each duty; but whenever the picquet marches from the head of the line, it passes for a duty both to the Officers and Soldiers, and is allowed them in their next tour.

If the picquet of one Regiment, or one Brigade, or that of one wing, should march, and not the rest, those Regiments whose picquet marched, are to be allowed it in the grand detail of duty.

Besides the Officers of the picquet already mentioned, there are General-Officers and Field-Officers appointed to command them.

The Generals so ordered, are called General-Officers of the day; and the Field-Officers are called Field-Officers of the picquet.

The General-Officers of the day for the infantry, are two, a Lieutenant-General, and a Major-General, who are to march with, and take the command of the picquet when it is ordered out upon any occasion: and as the picquet is immediately under the command of the Lieutenant-General of the day, it is not to march without his orders; therefore all orders relating to the picquet are sent directly to him, that he may give direction for its marching.

S   In

In cafe of an alarum, the picquets are to draw out at the head of their Regiments; but not to march from thence till they receive orders for it from the Lieutenant-General of the day; and though it fhould prove a falfe alarum, they are not to return to their tents till he orders them.

The number of Field-Officers appointed for the picquet, is according to the ftrength of the army. But, in order to give a proper idea of the ufual method, we will fuppofe a body of infantry, confifting of one hundred and eight Battalions, encamped in two lines; the front line confifting of fifty-fix Battalions, and the fecond of fifty-two. In this cafe, they always divide them into two bodies, diftinguifhed by the right and left wings. In dividing them, they do not feparate the bodies from one another, or leave a greater interval between the Regiments than ordinary; but only place the half on the right of both lines in the right wing, and the half on the left of both lines in the left wing; for the clearer underftanding of which, I have hereunto annexed a plan of the faid number of Battalions, divided into wings and Brigades, to fhew how the Field-Officers are appointed for the picquet; the number of whom to command the picquet of this body of foot, cannot be lefs, in my opinion, than four Colonels, four Lieutenant-Colonels, and four Majors, two of each rank for each line; by which each Colonel in the front line will have the command of the picquets of twenty-eight Battalions, which is fourteen hundred men, befides Officers, Serjeants, and Drummers: and thofe of the rear line will have the command of the picquets of twenty-fix Battalions, which is thirteen hundred men, befides Officers, &c. and though thefe numbers are above the ordinary commands of Colonels, yet, on extraordinary occafions, it is ufual for Colonels to have the command of fifteen hundred men; but more particularly fo in relation to the picquet.

When the infantry are thus divided into wings, they generally do duty apart; fo that each wing has a diftinct rofter, or roll of duty, kept for it. In this cafe, each

PLAN of 108 Battalions drawn up in Two Lines, divided into Wings & Brigades, to shew how the Field Officers are appointed for the Picquets.

Front Line, consisting of 56 Battalions, Form'd into 12 Brigades.

Left Wing of the Front Line, containing 28 Battalions, Forming six Brigades.   Right Wing of the Front Line, containing 28 Battalions, form'd into six Brigades.

Second Line, consisting of 52 Battalions, Form'd into 12 Brigades.

Left Wing of the Second Line, containing 26 Battalions.   Right Wing of the Second Line, containing 26 Battalions.

The Figures shew the Number of Battalions in each Brigade

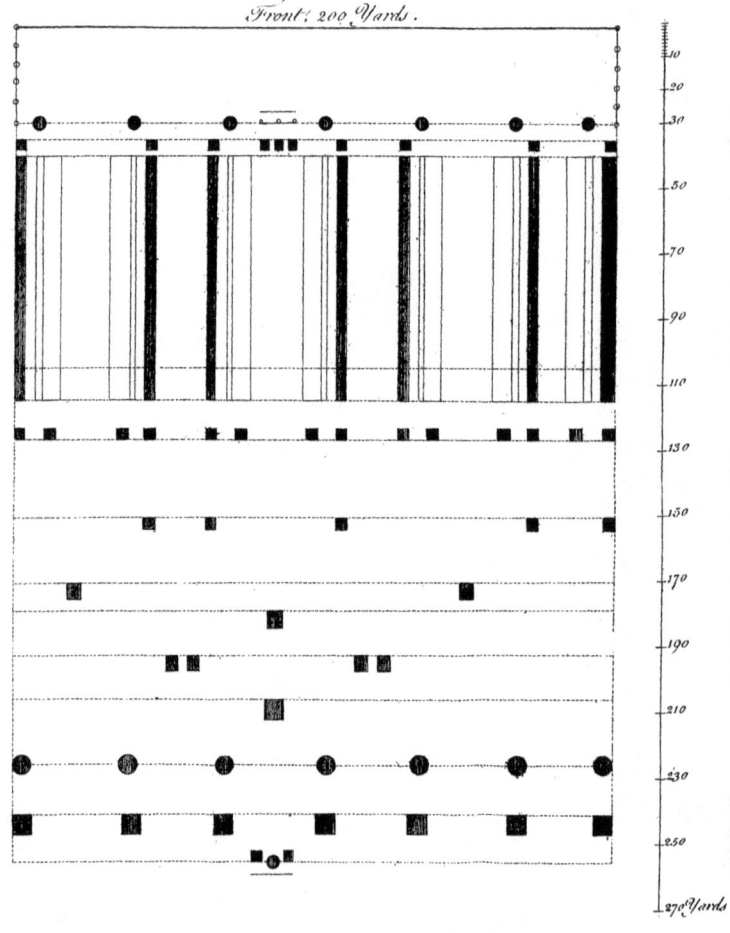

Plan of the Encampment of a Regiment of Dragoons, consisting of six Troops, forming three Squadrons: with the Light Troop on the right.

Chap. XV. *Military Discipline.* 259

wing furnishes its own Field-Officers for the picquet, and are appointed by name to their several commands in public orders in the following manner:

Such a Colonel, Lieutenant-Colonel, and Major, are for the picquet of the front line in the right wing.

Such a Colonel, Lieutenant-Colonel, and Major, are for the picquet of the second line in the right wing.

Such a Colonel, Lieutenant-Colonel, and Major, are for the picquet of the front line in the left wing.

Such a Colonel, Lieutenant-Colonel, and Major, are for the picquet of the second line in the left wing.

By the above method of appointing the Field-Officers, they can immediately repair to their several commands, and join the picquets, over whom they are placed, into a body, when they are ordered to march; and likewise know what quarter-guards they are to visit in going their rounds.

As the number over whom each Colonel of the picquet is placed, would be too great to be continued in one body, if they were ordered out upon service; every Colonel should therefore divide his men into two bodies, giving the command of the second to the Lieutenant-Colonel of the picquet who is under him; by which the picquet will be divided into eight bodies, four of which will consist of seven hundred men each, and the other four of six hundred and fifty men each, besides Officers, Serjeants, and Drummers.

The dividing the picquet in this manner, is not with a view to lessen the command of the Colonels, since those, over whom their Lieutenant-Colonels are placed, are to be still under their direction; but only to make them more fit for action, by reducing each body pretty near the complement of a Battalion; which model is certainly the most perfect, since all the infantry of Europe are divided into Battalions of about six or seven hundred men each, and therefore sufficiently evinces the truth of its being so.

When the *Retreat* beats, the Field-Officers of the picquet are to repair to the head of their respective wings, with their fashes on, to see the picquets drawn out, where the Colonels will give their orders to the Lieutenant-Colonels, and Majors, about going their rounds, &c.

The Colonels of the picquet go the grand round, which is done about the same time, and in the same manner, as the grand-round in a garrison.

When the Lieutenant-Colonels and Majors are ordered by their Colonels to go rounds, which is generally after the grand-round has gone, they are to be received on the footing of common-rounds, and are therefore to give the word to the Officers on guard.

The Field-Officers of the picquet, in going their rounds, are only to visit the quarter-guards of their wing.

The Lieutenant-Colonels and Majors are to make their report to the Colonels of the picquet, that is, to those under whose command they are placed, at the time which they shall appoint, which is commonly in the morning; and the Colonels of the picquet are to make their report of the whole to the Lieutenant-General of the day, at the head quarters, at orderly time.

If the General-Officers of the day think proper to go rounds, they are always to be received as grand-rounds, though the grand-round should be made; and the Officers must give them the word: neither are they, nor the grand-round, obliged to dismount when they receive the word.

Though it is positively said, that the picquet shall not march from the head of their Regiments till they receive orders for it from the Lieutenant-General of the Day, yet it cannot be supposed, but that the Major-General of the day, or the Colonels of the picquet, may venture to march them upon an emergency, without waiting for his orders; otherwise the intent and design of the picquet, that of putting a stop to the enemy till the army can draw out, may be obstructed. For as the Lieutenant-General of the day cannot be in every place at a time, should the enemy appear on one flank while he he is at the other, or in the center, his orders might come too late for their marching to oppofe them. It cannot therefore be doubted but that their marching on fuch an occafion, without waiting for his orders, is not only excufable, but abfolutely neceffary: however, unlefs there is a real neceffity, no fubordinate Officer fhould prefume to order the picquet to march, but by his commands.

The Lieutenant-General of the day may order the picquet of any Battalion to draw out under arms, for him to fee them, during any time of the night; but the other General-Officers of the day, or the Field-Officers of the picquet, cannot do it by their own authority.

I muft beg leave to offer one thing of my own, relating to the picquet, which is this:

When the picquet draws out upon any alarum, the Colonels of the picquet fhould have directions to join the picquets of their wing immediately into a body, without waiting for the Lieutenant-General of the day's orders; by which means they will be ready to march from the head of the line as foon as they fhall receive his orders, which will fave a great deal of time, that will be loft if they are not to join till he orders them. For as the faving of an out-poft, or a foraging efcort, when attacked, depends on the diligence of the picquet, the method I propofe will contribute towards it in point of time.

I do not mean that the whole fhould join in one body; but that every Colonel fhould join thofe into a body over whom he is appointed, the place for which fhould be in the front of the center Battalion of his wing; fo that the picquet would be formed into as many bodies as there are Colonels ordered for the picquet: and if it fhould prove a falfe alarum, they will have but a fhort way to march back to their Regiments, and therefore cannot be looked on as a fatigue; but if it fhould prove a real one, the advantage of their being joined, is, in my opinion, fo plain, that it will not admit of an objection; for when the General of the day sends them orders to march, his Aids-de-Camp will have an occasion to deliver them only to the Colonels of the picquet, which may be done in a very short time; whereas, by the other way, they must stop at every Battalion, to give the Captain of the picquet orders where he is to march to; and even after that, they must make a halt, in order to be formed into distinct bodies under the command of the Field-Officers, before they enter upon action: so that by the Aid-de-Camp's stopping at every Battalion to deliver his orders, and their halting afterwards to form them into distinct bodies, a confiderable time must be lost; whereas by the method proposed, they will be ready to march and enter upon action upon the first order.

As the *Germans* and *French* do not only differ from us but also from one another, in some particulars relating to their picquet-guard, I believe the inserting them here will not be thought improper; since the knowing the method of different nations, may be of service to an Officer on several occasions.

ARTICLE VI.

The picquet of the *Imperialists* and *French* consists of the same number of Officers and Soldiers from each Battalion as is ordered for ours, and designed for the same use, that of having a body of men always ready, &c. But what we differ in from the *Germans*, is this, that our picquets remain in their tents all night, and theirs serve as an advanced guard to the army.

As soon as the *Retreat* has beat, and that the Officers have examined the Soldiers arms and ammunition, they march their picquets about eighty or a hundred yards in the front of their own quarter-guards, where they are to continue all night, placing Sentries in their front and on their flanks; but those of the front line post a Serjeant's guard about thirty or forty yards advanced, when they are

Chap. XV.   *Military Discipline.*   263

are near the enemy, otherwise not, to which they send frequent patroles.

If the rear of their army lies open, or that they are under any apprehension of the enemy's attempting to surprise them there, the picquet of the second line is then posted about the same distance in the rear of their Sutlers tents, facing outwards, and taking the same precautions as those in the front line, in posting of Sentries and Serjeants guards, &c.

By this means, they say, both the front and rear of their army is secured, during the night, from being insulted by an inconsiderable number of the enemy's troops; and though they should advance with a large body, the picquet would stop them so long, till the army would have time to get to their arms.

This is the reason which they give for their picquet-guards lying out every night, in the manner above mentioned; and if they did it only when there was a real occasion, the reason would be good; but as it is their constant practice, from the opening of the campaign to the end of it, it proceeds rather from custom than necessity; since their situation cannot be always such, as to be liable to a surprise; and when they are not under those circumstances, acts of supererogation, in my opinion, ought to be avoided, that the men may not be fatigued to no purpose.

I do not from hence condemn the method, but the wrong application. When the armies are near one another, or that you are inferior to the enemy, or particularly so in horse, whose motions are quick; or that you are exposed to their insults by the situation of your camp; the drawing out of the picquet, in the *German* manner, will not be only proper, but absolutely necessary: but when they are not under these circumstances, it is very imprudent to act as if they were. It shews that the commander in chief has only attained to the mechanical part of the service, but wants judgment to apply it in the proper season.

A General should be careful, but not diffident, lest it make a bad impression on those under his command: neither should he despise the enemy too much, lest, by fancying himself in a state of security, he should give them an advantage over him; so that the true medium lies between presumption and diffidence; that is, bold, but not rash; circumspect, but not diffident. No man can attain to this merely by art. It must be implanted by nature, and brought to maturity by experience. Thus was our late victorious General, the duke of *Marlborough*, formed, whose conduct in war may be imitated, but hardly equalled.

The Field-Officers of the picquet do not go their rounds to the quarter-guards, but to the picquet-guards; and not only the grand-round is obliged to give the Officer of each picquet-guard the word; but the General-Officers of the day must do it also, if they come to visit the picquet.

All the Field-Officers of the picquet go their rounds. The Major begins his about half an hour after it is dark. His chief business is like that of a Town-Major, to see if the Sentries are properly posted, and if not, to give directions for the doing it; as also to examine into their numbers, &c.

The Colonel goes the grand round about twelve o'clock; and the Lieutenant-Colonel about an hour before day.

One of the General Officers of the day seldom fails of going to the picquet of the front line just at or after the beating of the *Reveille*; not on the footing of a round, since no round is made after *Reveille*; but only to ask how all things passed in the night, and whether they discovered any thing or not; after which he goes to visit the grand guards of horse, and out-posts.

About half an hour after the *Reveille*, the picquet-guards return to their Regiments. This is the method of the picquet guards of foot of the *Imperialists*; and I presume it is the same amongst the troops of all the *German* princes.

The

The horse and dragoon picquets are formed also at the head of their respective wings; but do not advance above thirty paces from the standard-guards.

At the setting of the watch, they are drawn out at the head of their Regiments, as the foot are; afterwards those of each wing are joined, and then formed into squadrons, with proper Officers to command them; and as soon as they have posted their Videts, or Sentries, the men are ordered to dismount, and lie at their horses heads.

As they are allowed to bring forage with them to feed their horses, they have liberty to unbridle.

When they are under any apprehensions of the enemy's attempts, they do not only post Subalterns guards in their front and flanks, but keep one entire rank of the whole mounted, which they relieve every hour, by making the three ranks take it in their turn; and send frequent patroles round their Videts and advanced guards.

The Field-Officers of the horse-picquet assemble with the men, and remain with them all night.

A little before day, the men are ordered to bridle their horses, and mount, that they may be ready to sustain the grand guards, in case they are attacked; and a little after day-break they send out patroles to reconnoitre as far as the grand guards of their own wing, with orders for them not to return, till they are marched to their day-posts. They send out patroles likewise to reconnoitre on their flanks; that is, those of the right wing reconnoitre the right flank of the army; and the left wing, the left flank; which patroles seldom reconnoitre above a mile, unless on some extraordinary occasion. As soon as their patroles are returned, and their guards and Videts drawn in, the horse-picquets return to their Regiments.

The number of horse which mount the picquet, is not fixed, as it is in the foot; but are more or less according as there is occasion, both with them and us: however, there are never less than two of a troop, and seldom

more

more than eight; so that according to the number of men, Officers are commanded in proportion.

The common method is to appoint a Captain, Lieutenant, and Cornet, to eighty or a hundred horse; but sometimes they order the same number of Officers to fifty, sixty, or seventy, according to the service on which they are commanded. It is a standing rule in the cavalry, that when a Captain of horse is ordered on duty, he has a Quarter-Master and Trumpet from his own Regiment, so that they are never mentioned in the orders; but the Lieutenants and Cornets have neither, unless on very particular Occasions; such as a guard of horse on the General in Chief.

The manner of joining the picquet of each wing of horse into a body, and then forming them into Squadrons, is preferable to the method of their foot-picquet, that of posting the picquet of each Battalion single: for should the enemy make an attempt upon the encampment of the foot in the night, they could not be opposed at any one place of it with more than a Captain and fifty men, besides the quarter-guard, which could not make any considerable resistance, at least not sufficient to give the Regiments time to form as they ought. It is true that the foot are not very liable to surprises of this nature, from their being encamped in the center: however, I am of opinion, that when the picquet is to continue out all night, they should be formed into Colonels commands; that is, into as many bodies as there are Colonels appointed for the picquet, and the Field-Officers to remain with their several commands. After they are thus formed, the General of the Day should post them in such places along the front, flanks, or rear, by which the enemy can have no access to the camp; so that by having so considerable a body, as a Colonel's command, posted at each avenue leading to your camp, the army can be in no danger from a surprise, since they will be able to make such a resistance, as will give the army time to form, which is all that is required from

Chap. XV. *Military Discipline.* 267

the picquet, that being the end for which they are then designed.

When the rear is so secured that it is in no danger from the enemy, the picquet of that line may be posted likewise in the front of the first line, or on the flanks, if requisite.

From each of these bodies, the Colonels should post Officers guards at a proper distance in their front, and small ones on their flanks, and sends frequent patroles round them to keep them alert.

The Officers of these advanced guards must be extremely vigilant and watchful, that they may not be surprised; for which end they should post two Sentries at each post in their front, that one may come to the guard, upon the appearance of any number of men, or the hearing of a noise like the march of troops, to acquaint the Officer with it; upon which they should put their men under arms, and endeavour to inform themselves thoroughly of the truth of that report, by reconnoitring the place where the men were seen, or from whence the noise was heard, before they send an account of it to the Colonel of the picquet.

When the Officers find the report of the Sentries to be true, they are to send an account of it immediately to the Colonel of the picquet, and so from time to time as they discover any thing further.

If the number of men which were discovered, or the noise which was heard, is considerable, the Colonel of the picquet is to send an account of it to the General of the day, as also to those bodies which are posted near him, with directions for them to communicate it to those next them, and so from one to another, till the several bodies of the picquet are acquainted with it, that they may be all prepared for their defence, or ready to march upon the first order they shall receive from the Lieutenant-General of the day.

On such notice, the Colonels of the picquet are to put their men under arms, and to send patroles round their advanced guards and Sentries, to see that they are alert,

and

and to acquaint the Officers who command those guards, with the report which was sent them from the other bodies, that they may be ready to oppose any attempt that shall be made on them; as also to reconnoitre beyond their advanced Sentries, and to send him a report of what they discover.

As the Colonels of the picquet cannot quit their posts to sustain one another, without they have orders for it from the General of the day, it cannot therefore be doubted, but that he will immediately repair to the place from whence he received the report, in order to give such directions as the service may require: for which end the General-Officers of the day should have a tent pitched in the rear of that body, which is posted opposite to the center of the first line, where they should remain all night, that the Colonels of the picquet may send their reports when any thing is discovered or heard, and receive their commands, without loss of time.

Having given full directions in the 6th Article, Chapter XI. how Sentries are to behave on their posts, there is no occasion for its being further mentioned; but lest the enemy should advance upon the advanced guards before the General of the day can arrive at that post to give the necessary orders, they are not to quit their posts till they are forced to it by superior numbers; and even in that case they are to maintain it as long as they can by firing upon them, after which they are to retire slowly to the body from which they were detached.

Upon the fire of the advanced guards of foot, the horse picquet should immediately mount, and the General-Officer who commands them should send a sufficient detachment towards the place from whence the fire was heard, in order to sustain the post that is attacked; and be ready to follow with the whole, in case those should not be sufficient: however, all the horse picquet should not march from their posts, till part of the cavalry are mounted and formed, lest the enemy should have done it with a design to draw them from thence, and then attack

tack the flanks of the army before they are prepared to receive them. But I muſt not proceed further on this head, for fear of incurring the juſt cenſure of my ſuperiors, by preſuming to lay down rules to thoſe who are thoroughly verſed in all things relating to the ſervice. I hope they will therefore excuſe the liberty I have taken in entering, perhaps, a little too freely into the grand detail, which I was neceſſitated to do now and then, or leave the parts treated on not clearly underſtood by thoſe for whom it is deſigned, young Officers. Beſides, as there is great reaſon to believe, that ſeveral of them will arrive to the rank of General Officers, before they have an opportunity of acquiring a knowledge of their duty by ſervice, thoſe things relating to it, which are here inſerted, I believe, upon ſecond thoughts, will not be judged altogether improper, in order to give them a ſmall idea of thoſe important poſts.

When the horſe and foot picquet is diſpoſed of according to the above method, the army can be under no apprehenſions of a ſurpriſe. Beſides, enterpriſes of that nature are ſeldom ſuccefsful, from the difficulties that attend night expeditions, and therefore very ſeldom undertaken; but if they ſhould attempt it, and paſs the grand-guards of horſe without being diſcovered, which is not very eaſy if they perform their duty, the oppoſition which they will meet with from the picquet, both of horſe and foot, will, in all probability, make them conclude that their deſign is diſcovered, and conſequently give it over, and return from whence they came; but if it has not that effect, it will however give the army the time requiſite to draw out and oppoſe them.

I ſhall now proceed to the method of the foot picquet of the *French*.

The picquet-guard of each Battalion conſiſts of the ſame number of Officers and Soldiers, and deſigned for the ſame end as ours, with this difference; that they furniſh Sentries to their Colours and Bells of arms, their quarter-guard being compofed of a Serjeant and twelve men only.

The picquet is drawn up in the interval on the right of the Grenadiers, where they continue till they are relieved at *Retreat-beating*; for which reafon, they always erect a fhed there, made with boughs and ftraw, to keep their arms and ammunition from the rain.

When the Commander in Chief of the army, or the General Officers of the day, pafs by, they only draw up in their ranks without arms; and if the King fhould pafs by, they pay him no other compliment.

The day the army is to march, at the beating of the *General*, the Officers of the picquet are to get on horfeback, and to take care that the Soldiers do not take their arms out of the bells, and go before, or ftir from the Battalion, which the *French* Soldiers would frequently do, were it not for this precaution.

When the Regiment is drawn out, either to march, or to mount the trenches the picquet is always formed on the right of the Battalion, and marches immediately after the Grenadiers.

If a Captain and fifty men of a Regiment are commanded out, while they are on the march, or in trenches, the picquet is to perform that fervice; that is, they are to march firft; but if a fecond detachment of the fame number is wanted, it is taken from the Battalion, the firft having finifhed their duty; otherwife, the firft picquet would pafs their time extremely ill in the trenches, were they to be commanded out before the others have taken their tour. This is all in which the *German* and *French* picquets differ from ours.

CHAP.

CHAP. XVI.

Confiſting of the guards ordinary of the Horſe and Dragoons; and alſo extraordinary guards of the Foot.

ARTICLE I.

FORMERLY the horſe and dragoons were looked upon as two diſtinct bodies, and therefore had ſeparate duties: for the horſe did all the duty of the grand-guard, and the dragoons that of convoys or eſcorts, guarding of paſſes and fords, as being rather expeditious foot, than horſe; for which ſervice they were more uſeful than for field action, their horſes being too ſmall to ſtand a charge: But, in the late war they were ſo well mounted, that they rolled in all duties with the horſe, and therefore compoſed but one body, under the denomination of cavalry.

The guards ordinary of the cavalry, are the ſtandard-guards, and grand-guards.

Each Regiment has a ſtandard-guard, which is of the ſame nature with the quarter-guards of the infantry, that of a guard to the Regiment, and to pay the compliment due to the general Officers.

The ſtandard-guard for a Regiment of horſe, commonly conſiſts of a Corporal and twelve Troopers; but, at the ſetting of the watch, they have ſix men added to them, they having more Sentries by night than in the day-time, which additional men go off ſoon after *Reveille*.

The ſtandard-guard of a Regiment of dragoons, conſiſts of the ſame number, with the addition of a Serjeant.

The

The standard-guards are relieved every morning at the beating of the *Troop*, and the men mount on foot, and are drawn up on each side of the standards, in a single rank, facing outward. They have neither Trumpet nor Drum; so that they can pay no other compliment to the Generals, than that of the horse resting their carbines on their left arms, and the dragoons resting their firelocks as the foot do.

In the day-time, the men of the standard-guard are obliged to appear in boots; but at night they throw them off: which custom of mounting in boots on the standard-guard, is certainly ridiculous, since there is no end proposed by it: for as the men of the standard-guard are not to leave the camp, unless the Regiment is ordered to march, to what purpose are they to have their boots on? neither are their horses saddled at the picquet; therefore there can be no reason given for it, that I could hear of, but custom; which is but a poor support for what in itself is both inconvenient and absurd. Besides, to order them to mount a guard on foot in their boots, and their horses unsaddled at the picquet, appears so inconsistent, that I am surprised it has not been abolished long since. Where a custom is of a long standing, and though there can be no great use made of it, yet if it is not attended with any inconveniency, it may be continued as a thing indifferent, but not otherwise: therefore I presume, as the gentlemen of the cavalry have entered a little further into the spirit of discipline than formerly, they will lay aside the custom of making the men of the standard-guard mount in boots.

ARTICLE II.

The grand-guards are done by detachment, and are relieved every morning at the same time that the standard-guards are.

When the army is large, each wing of horse does duty by itself, without intermixing with one another, and therefore have distinct grand-guards allotted them; for which

which reason they have separate parades. That for both lines of the right-wing, is generally about the center of the front line of horse on the right; and that for both lines of the left-wing, opposite to the center of the front line on the left; on which parades those who are ordered for the grand-guards are to assemble, and to be detached from thence by the Majors of Brigade.

The grand-guards are divided into Captains commands, in each of which there are seldom less than fifty men, or more than an hundred, and each Captain has a Lieutenant and Cornet along with him.

The number that mounts daily is not fixed; but depends on the number of your troops, the situation of your camp, or the neighbourhood of the enemy; according to which there are more or less ordered: however, the common rule is to post a Captain's command at, or near each avenue in the front of the army, by which the enemy can approach the camp, unless they should lie in low grounds or bottoms, the eminences being the properest places to post them on, that they may discover the march of troops a good way off, and give notice to the camp of the approach of the enemy; as also to keep off small parties from plundering or molesting it.

When each of these guards consists of eighty or a hundred men, they generally post a Lieutenant and thirty troopers, or a Cornet and twenty, at a proper distance in their front, but not out of view, to give them notice when any party appears; but when they only consist of fifty or sixty men, they seldom detach to these advanced guards above a Quarter-Master and sixteen troopers, or a Corporal and twelve.

The grand-guards should never be posted in a narrow pass or road, or too near a wood, but at some distance from them, lest they should be surprised by a party of foot, or partisan parties, which generally lie lurking there; but when such places lie near their posts, they should place Videts or Sentries pretty near those roads or woods, to give the guard notice to mount when any number of armed men appears; they should likewise send small patroles to reconnoitre those places frequently, otherwise they may be surprised and carried off, when they least think of it, by an inconsiderable number of men.

The Videts which are posted in the front, or near those suspected places, should be placed double, that one may come and acquaint the Officer of the guard when they discover any body of men, and the other to remain at his post till the enemy advance upon him, and force him from thence, or endeavour to cut off his retreat, by getting betwixt him and his guard; on either of which, he is to fire his carbine, (which all Videts are to keep advanced upon their right thighs for that purpose) and return to his guard; but unless for the reasons just mentioned, no Videt is to leave his post till he is regularly relieved by the Corporal of the guard.

The grand-guards keep their front always towards the enemy; neither do they change it when the Generals come to visit them: however, when any of them come, the grand-grands are to mount, and receive them with drawn swords and sound of trumpet; for which reason, they should always have a Videt betwixt them and the camp, to give them notice of the approach of the Generals, that they may have time to mount, and pay the compliment due to them.

When the grand-guards are relieved, they do not draw up opposite to one another, as the foot do; but the new guard draws up on the left of the old one, if the ground will allow of it, otherwise in the rear of it; and as soon as the old guard is marched off, the new guard draws up on their ground.

The grand-guards have two posts, one by day, and another by night.

The day-post is sometimes a mile from the camp, or more or less, according to the situation of the ground, or the vicinity of the enemy; it being necessary to post them in such places as will admit of a view that they may discover a good way into the country.

The night-poſt is generally within half a mile of the camp, to which they retire at the ſetting of the watch, to prevent their being carried off in the night by the enemy, the day-poſt being at too great a diſtance to remain there with any ſafety, ſince the picquet could not come time enough to their aſſiſtance, ſhould they be attacked; but by their drawing near the camp at night, the enemy cannot ſo eaſily inſult them; or, if they attempt it, they can be immediately ſuſtained by the horſe and foot picquet.

In order to put it more out of the power of the enemy, the night-poſts are frequently changed, and new ones aſſigned them every third or fourth night, or oftner if there is occaſion; by which means the enemy cannot be ſure of the place they are poſted at, and will therefore render their attempt very uncertain.

Immediately after the *Reveille* has beat, the grand-guards march to their day-poſts, and ſend ſmall parties a little before them to reconnoitre all ſuſpected places, to avoid falling into an ambuſcade; which they might eaſily do without this precaution, ſince they ſeldom march to their day-poſts, but that they diſcover a party of the enemy's horſe, or huſſars, at or near the ſaid poſts.

The Lieutenant-Generals of the day are the proper Officers to whom the poſting of the grand-guards belong; and after they are poſted by them, none but the Commander in Chief of the army, and the Generals of horſe and foot, have a power of altering them.

The men of the grand-guard always carry forage with them to feed their horſes: however, I preſume, they are never allowed to unbridle the whole at the ſame time, but that one rank is always kept ready to mount, unleſs they are ſo advantageouſly poſted, that they command the view of the whole country.

During the night, they always keep one rank mounted, which they relieve time-about, and ſend frequent patroles round their advanced guards and Videts, to keep them alert.

A little before day, they all mount, and continue so till they march to their day-posts.

When any of the grand-guard discover any number of men, whether horse or foot, they are to mount immediately, and to send out a Corporal and four or six troopers, who are well mounted, to reconnoitre them near, in order to discover whether they are friends or foes, and their numbers; and when it proves to be the enemy, and that their numbers are considerable, they are to send an account of it immediately to the General of the day, that he may order the picquet to draw out, that they may be ready to oppose them, in case they should attack the grand-guard.

The Officer who commands that part of the grand-guard from which the enemy was discovered, should likewise send an account of it to those detachments which are posted near him, who are to send the same account to the next, and so from one to another, that they may all prepare for their defence.

The grand-guards are not to quit their posts till obliged to it by superior numbers; and even in that case, they are not to go off with precipitation, but retire in a slow and regular manner before them, and to dispute every spot of ground that will admit of it, in order to put a stop to them till the picquet can be brought to oppose them.

When there are several Captains ordered for the grand-guard of each wing, Field-Officers, in proportion to the number who mount, are appointed to command them: in which case all reports, from the several detachments of the grand-guard, relating to the discovery of the enemy, &c are to be made to them, and they to the General of the day: and according to the disposition of the enemy, the Field-Officer who commands the grand-guard may join the whole into one or more bodies, as he shall judge proper for the service; without which power, the detachments may be attacked and beat one after another, who when joined may be sufficient

cient to repulse the enemy, or put a stop to their progress till the picquet can come to their assistance.

As soon as the grand-guard is relieved, the Officer who commands it is to make his report to the Lieutenant-General of the day.

ARTICLE III.

Guards Extraordinary.

By this is meant those guards, or detachments, which are only commanded on particular occasions; either for the further security of the camp, which are called Out-Posts, or to cover the foragers of the army, for convoys or escorts, or for expeditions; so that the proper term is rather Extraordinary Commands or Detachments.

These commands, by what denomination soever called, are done by detachments; and each Battalion, whether strong or weak, furnishes an equal proportion of private men to them.

OUT-POSTS.

When a body of men are posted beyond the grand-guard, they are called Out-posts, as being without the rounds, or limits of the camp.

The occasion of their being commanded, is generally to prevent the army from being surprised, or disturbed in the night by the enemy, or to secure a pass or ford on a river, or village, or villages, that may lie between the two armies; as also to keep a communication open with your own garrisons, or to cover your convoys of provisions, to prevent their being annoyed by the enemy.

All the out-posts which lie near the camp are relieved every morning with the guards ordinary; but those which are at any great distance, such as three, four, or five miles, are generally relieved but once in four or eight days.

The same directions which are given for parading the men for the guards ordinary, must be observed in parading of those for the out-posts, with this addition, that the Adjutants must see that the men who are to continue any time on duty, are sufficiently provided with ammunition bread and pay.

When the out-guards are posted in villages, they should strengthen themselves in them as much as possible, by throwing of barricades cross each street, or entrance into them; but when the entrances are too many to be defended any time, they should likewise strengthen the church-yard, or any other part of the village, which they find more proper for their purpose, to retire to when they are forced from the others, that they may be able to defend themselves till relieved by their army; but when an out-post has not the conveniency of a village, church-yard or house, a fort, composed of fascines and earth, should be thrown up to secure them, which may be done in a very short time; otherwise the detachment may be carried off any night by the enemy.

When the out-posts which lie near the camp have relieved, they are to send an orderly man from each to attend at the Major of Brigade's tent of the day, in order to conduct the guards which are sent to relieve them; as also to carry what orders the Major of Brigade of the day shall receive for those posts, from their time of mounting till they are relieved; after the delivery of which, the orderly men are to return to the Major of Brigade's tent, and acquaint him of their having delivered them. These orders should always be sent in writing, and sealed up, lest any mistake should happen through the negligence or wrong construction of the orderly men; as also that the Officers, who command those posts, may be able to justify their conduct, by producing the said orders, in case the obeying them should be attended with any ill consequence.

The Officers should take particular care to send such men orderly, whose fidelity and sobriety they can rely on most.

The

The out-pofts are to turn out, and receive the Generals, who come to vifit them, under arms; but not to beat a Drum, though the commander in chief of the army fhould come to vifit their pofts.

The out-pofts which are near the camp are to have the parole which is given to the army, fent them in writing by their orderly men; but thofe who are at a diftance fhould have a parole and counter-fign of their own fent by an orderly Trooper; the care of which belongs properly to the Adjutant-General of the army, as thofe which lie near the camp do to the Major of Brigade of the day.

The commanding Officer at each out-poft is to fee his night Sentries pofted before it is dark, and at the advanced pofts he is to place them double, for the reafons already premifed.

During the night, the Sentries at the out-pofts fhould be relieved every hour; and between every relief a patrole fhould be fent round them to keep them alert; fo that by the relief and the patrole all the pofts will be vifited every half hour. My reafon for this, is not only to keep them very watchful, but likewife to prevent the ill confequences that may attend their deferting to the enemy, or quitting their pofts, fince they cannot be gone long before it is found out; and as often as the Sentry is miffing, the Officer who commands the out-poft is immediately to change his counter-fign, and fend it to all his Sentries; for fhould the Sentry who is miffing defert to the enemy, and difcover the counter-fign, they *might impofe on your Sentries, and furprife the guard*; but by their being vifited fo often, it will be found out before they can poffibly have time enough to execute the defign, unlefs your poft lies very near the enemy, in which cafe it is requifite for the whole to be as alert as the Sentries. It is therefore incumbent on the Officers who command out-pofts to be very exact in this part of their duty, or they and their parties may be eafily deftroyed by the treachery of a Sentry.

When they are obliged to change their counter-sign for the above reason, they should send an account of it immediately to all guards, or out-posts, with whom they have a communication, that they may do the same, lest the enemy should attempt to surprise them.

As the safety of an army may often depend on the out-posts, the Officers who command them cannot be too exact in the discharge of their duty; they ought therefore to be very vigilant, and not think giving the necessary orders sufficient, but see them executed also; otherwise they may be deceived by trusting entirely to reports. It is on these commands where Officers have frequent opportunities of distinguishing themselves: it is therefore to be presumed, that whoever has a regard to his reputation or fortune, will not be so much wanting to himself, as to neglect the common rules which are here laid down for his conduct.

The Officers who command out-posts, should order their men to stand to their arms a little before break of day, and to continue so till it is so light that they can see a mile or two from them, it being usual for troops to advance near a post in the night, but defer attacking it till they can distinguish one man from another, for fear of destroying their own instead of the others; besides, as the morning is the time that every man is most sleepy, it is therefore the more necessary to use this precaution, in order to have them thoroughly awake, that they may be the better prepared for action, in case of an attempt.

How far an Officer who commands an out-post, should persevere in the maintaining of it against a superior body of troops, cannot be declared, without knowing both his orders and situation; but though they should be general, he ought not to quit it, if there is a probability of his maintaining it till he can be relieved by his own army, unless he has orders to retire upon the approach of a superior force; but if his orders are positive, and directs him to defend it to the last man, he must obey them, even against a whole army, without reflecting on the

the consequences. But such orders as these are never given, unless the preservation of your army, or the country, depends on it: for as the custom of war is otherwise, it would be deemed madness, and not bravery, for a party of men to pretend to defend themselves in a village, house, or church, or any place that is not tolerably well fortified, against an army, when they cannot be supported by their own troops, but must be taken when attacked: but when an Officer is posted in a place that cannot be taken without Cannon, he is not to surrender it till he is regularly attacked and a breach made, or the place so battered, that it is no longer tenable, let them send ever so many threatening summons of hanging, or putting all to the sword if they do not, since the rules of war do not authorise such pieces of cruelty. Besides, a generous enemy will be so far from committing it, that they will esteem and value him for his behaviour, if he does not persevere beyond what a prudent and brave man ought; whereas, should he surrender before he is reduced to a necessity of yielding, they will look upon him as a man void of courage and conduct, and despise him as one, whose fear had betrayed him into an unworthy action; and if an Officer is despised by the enemy, for his ill conduct, as he certainly will, he surely deserves the highest punishment from his friends for it.

ARTICLE IV.

Foraging-Parties.

These parties are to secure the foragers from being taken by the enemy, or disturbed while they are foraging.

According to the danger which your foragers may run, by the place they are to forage in being near to, or remote from the enemy, the covering parties are stronger or weaker.

In inclosed countries, the covering parties consist for the most part of foot; but in a champaign country, they are generally composed both of horse and foot. These detachments march generally from the camp the night before the army is to forage, in order to possess themselves of the posts which they are to guard, before the foragers leave the camp; and as soon as all the foragers have got their forage, and returned with it to the camp, the covering party does the same.

When the army is large, or that they lie near the enemy, they seldom suffer the whole army to forage the same day; but order one wing to forage one day, and the other wing another day; in which case, the wing which forages sends detachments to cover their own foragers; neither should they be allowed to send above three men of a tent from the cavalry to forage at a time, that they may have a sufficient number to defend the camp till the foragers return: but lest the enemy should take the advantage of your foraging, and endeavour to attack your camp in their absence, upon the first notice of their march, the General orders the signal to be made for the foragers to return, which is generally the firing of three pieces of cannon; on the hearing of which the foragers are to leave their forage and repair immediately to their Regiments, and the covering parties are to return likewise to the camp.

As these detachments are posted between the enemy and the foragers, they are not to suffer any of the foragers to pass beyond them, in search of forage, lest they should be taken; which danger they would always run, without reflecting on the consequence, were they not detained from it by the covering parties: it is therefore the duty of the Officers on these commands to prevent their doing of it, and to compel them by force to keep within the bounds prescribed them.

ARTI-

ARTICLE V.

Convoys or Escorts.

These are to conduct the bread waggons and other provisions; dry forage, at the opening or closing of the campaign; ammunition, heavy cannon or field-pieces; as also persons of distinction who are coming or going from the camp.

The convoys are generally done by detachment; but when they are to pass near the enemy's garrisons, or liable to be intercepted by a considerable body of their troops, it is usual to command entire Brigades both of horse and foot on that service; or in lieu of Regiments of foot, a sufficient number of Companies of Grenadiers, for the greater expedition.

ARTICLE VI.

Expeditions.

These parties are sent into the territories belonging to or under the protection of the enemy, to destroy the country, or lay it under contribution; as also to intercept their convoys, and straiten them in their camp: but as these parties cannot remain long in a place, lest the enemy should fall upon them, they are generally composed of cavalry, the infantry not being expeditious enough for that sort of service.

They are likewise sent to fall upon the enemy's foragers even in the rear of their camp; but as this is attended with a great deal of danger and difficulty, it is very seldom undertaken.

Formerly these sort of exploits were very much in vogue, particularly with the *French*, who call it *La Petite Guerre*; but of late they are very much left off, since they only serve to render the poor inhabitants more miserable,

miserable, or particular Officers, whose horses or baggage they take, uneasy in their affairs, without contributing any thing to the service, or the bringing of the war the sooner to a conclusion. Besides, by the great fatigue which it brings on your own troops, a great many horses will not only be rendered unfit for immediate service, but entirely lost; which reason is sufficient, in my opinion, to discontinue the practice, at least not to use it but on particular occasions.

CHAP. XVII.

General rules for the encamping of an army, with the particulars for the encamping of a Regiment of horse, and a Battalion of foot; and two plans of the same.

ARTICLE I.

Proportions to be observed, in encamping a Regiment of Dragoons, of six Troops, forming three Squadrons; with the light Troop.

FRONT two hundred yards, divided as follows:

Yards

For pitching six rows of tents, with the intervals between the tents and the pickets, length for the standing of the horses, and space for laying up the dung, at fourteen yards each — 84

For the breadth of three streets, between the horses of each Squadron, at seventeen yards each — 51

For the breadth of two back-streets, at twenty yards each — 40

For the breadth of one street, between the first Squadron, and the light Troop — 9

For pitching the tents, &c. of the light Troop, as above — 16

Total front 200

The

The fourteen yards allowed for the front of each troop, is divided os follows :

	Yards
For pitching a horseman's tent	3
From front-pole of the tents, to the pickets	3
From the pickets, to the edge of the dung	6
Breadth of the dung	2
	14

N. B. The sixteen yards allowed for the front of the light Troop, are divided in the same manner; the two additional yards being allowed for pitching their tents, which are larger than those of the other troops.

	Yards
The interval between two Regiments of dragoons, is	60
Total front, and interval	260

Depth 258 yards, divided as follows :

	Yards
From the first line of parade, to the bells of arms	30
From the bells of arms, to the front-poles of the Quarter-Masters tents	5
From the front-poles of the Quarter-Masters tents to the first picket	5
Allowed for the standing of sixty-six horses	66
For the standing of the Subalterns horses, which are in a line with the troop	10
From the rear of the Subalterns horses, to the front of the Subalterns tents	12
From the front of the Subalterns tents, to the front of the Captains	24

From

Chap. XVII.   *Military Discipline.*   287

	Yards
From the front of the Captains, to the front of the Field-Officers	20
From the front of the Field-Officers, to the front of the Colonel's	8
From the front of the Colonel's, to the front of the Staff-Officers	14
From the front of the Staff-Officers, to the front of the grand-sutler	14
From the front of the grand-sutler, to the center of the kitchens	20
From the center of the kitchens, to the front of the petty sutlers	15
From the front of the petty sutlers, to the center of the bell of arms, of the rear-guard	15
Total depth	258

The parade of the standard-guard is four yards advanced before the center of the bells of arms.

The bells of arms are in a line with the pickets.

The standard-guard tents are pitched in the center of the third squadron, in a line with the fronts of the Quarter Masters tents, and are three yards distant from center to center.

Eleven tents are pitched for the men of each troop: the centers of the first and last, are three yards distant from the ends of the pickets: the others are six yards distant from center to center.

Seven tents are pitched for the light-Troop: the centers of the first and last, are fourteen feet one inch and a half from the ends of the pickets: the others are twenty-eight feet three inches from center to center.

The dung of each troop is laid up behind the horses.

The sixty-six yards are divided into four spaces of fifteen yards each, with three intervals of two yards each.

The

The dung of the Subalterns horses, is laid up in the space of eight yards; leaving an interval of two yards between it, and that of the troop.

The Subalterns servants front their horses.

The Lieutenant-Colonel's tent fronts the center of the first squadron.

The Major's tent fronts the center of the second.

The Colonel's tent fronts the standards.

The Staff Officers front the two back-streets on the right and left of the center, or third squadron.

All the Officers, the Subalterns excepted, have their horses in the rears of their tents.

The grand-sutler is placed in the rear of the Colonel.

The centers of the kitchens, are in the lines of the pickets produced: the inner diameter is sixteen feet; the breadth of the trench surrounding them, is three feet; the seat is one foot and a half; and the breadth of the outside wall two feet; which makes the outer diameter twenty-nine feet.

The front poles of the petty sutlers tents, or huts, are in a line with the centers of the kitchens, allowing to each petty sutler six yards in front, and eight in depth, to be enclosed by a trench one foot in breadth, and the earth thrown inwards.

The front poles of the rear-guard tents, are in a line with the center of their bell of arms; and distant from each other six yards.

The rear-guard fronts outwards.

The parade of the rear-guard is four yards distant from their bell of arms.

As it is usual for the Subalterns of horse to have a tent each, I have therefore placed two in the rear of each troop; and, though by the former method of encamping, they were generally pitched in a line with the troopers tents, and faced towards the streets as they did, yet, in this plan, I have placed them according to the manner of the foot, by facing them towards the Captains tents with a street of twenty-four paces between them.

The

The dimensions of the Captains and Subalterns tents, are as follows.

	Feet.	Inch.
Lenth of the ridge pole	7	8
Height of the standard poles	8	
Length from front to rear between the half-walls of the marquise	14	
Breadth of the marquise between the half-walls	10	6
Height of the half-walls of a marquise	4	

The Lieutenant-Colonel's, and Major's tents, about a foot larger.

The ornaments of all Officers tents to be uniform, and anfwerable to the facings of the Regiments they belong to.

The size of the troop-tents is sufficient for those of the Quarter-masters; only that they are allowed to have them a little higher, with a small marquise to throw off the rain.

The dimensions here given for the Officers tents, may be thought by some too small; and if they were only to encamp in *Hide-Park*, I should be of the same opinion; but let those gentlemen who think so, only make one real campaign, and I am convinced, they will wish them rather of a less size than a greater.

The circles which are drawn in the plan between the grand and petty sutlers, are marked for the kitchens, or places where the private men are to dress their victuals: they are made in the following manner:

First, you draw a circle or square on the ground of the dimension above limited, after that you dig a trench or ditch round it of three feet broad, and two deep, by which it will resemble the bottom of a cock-pit. When this is done, you are then to cut holes or niches in the side of the circle or square of earth which is left standing within the ditch. These holes may be about a foot square, the upper part of which should be within three or four inches of the surface, from whence they are to cut small holes of four inches diameter, down to the great ones, in which the fire is to be made, and the heat conveyed through those small holes to the bottom of the kettles, which are placed on the top of them. These fire-places may be made within three or four feet of one another quite round the said circle or square; and if you erect one of these kitchens (by which I mean an entire circle or square) for each troop or company, they need not be larger than what will contain as many fire-places as you have tents pitched for your troop or company; for as all the men who lie in a tent, are of one mess, every mess must therefore have a fire place, that they may have no excuse for their not boiling the pot every day.

There are several advantages by making of the kitchens as here directed.

First, A very little fuel will serve to dress their victuals; for as the fire-places are open at the side, like the mouth of an oven, the air which enters there forces all the heat up the small hole to the bottom of the kettle, and consequently boils it very soon; and, as the kettle covers the said hole, the rain cannot come to extinguish it, or create the men any trouble in keeping of it in.

Secondly, They are not in great danger of accidents by the fire's being blown amongst the tents or forage; for, if the men only lay a sod or turf on the top of the hole when they take off the kettle, it cannot be dispersed by the wind, which, without this consideration, they ought to do, in order to keep the fire-places dry.

Thirdly, The cutting of a ditch round the kitchens, does not only enable them to make the fire-places, but likewise prevents the fire from catching hold of the stubble or grass, which, in very dry or hot weather, it is apt to do, and endanger the burning of your camp, which I have often seen for the want of this precaution. Besides, the opposite bank of the ditch serves as a seat for the men who are employed in dressing the victuals.

Fourthly and lastly, By having of kitchens made in this manner, the Officers can, with a great deal of ease, look into the conduct and œconomy of their men, and oblige

the

Plan of the Encampment of a Battalion of Foot of 9 Companies each consisting of 70 private Men.

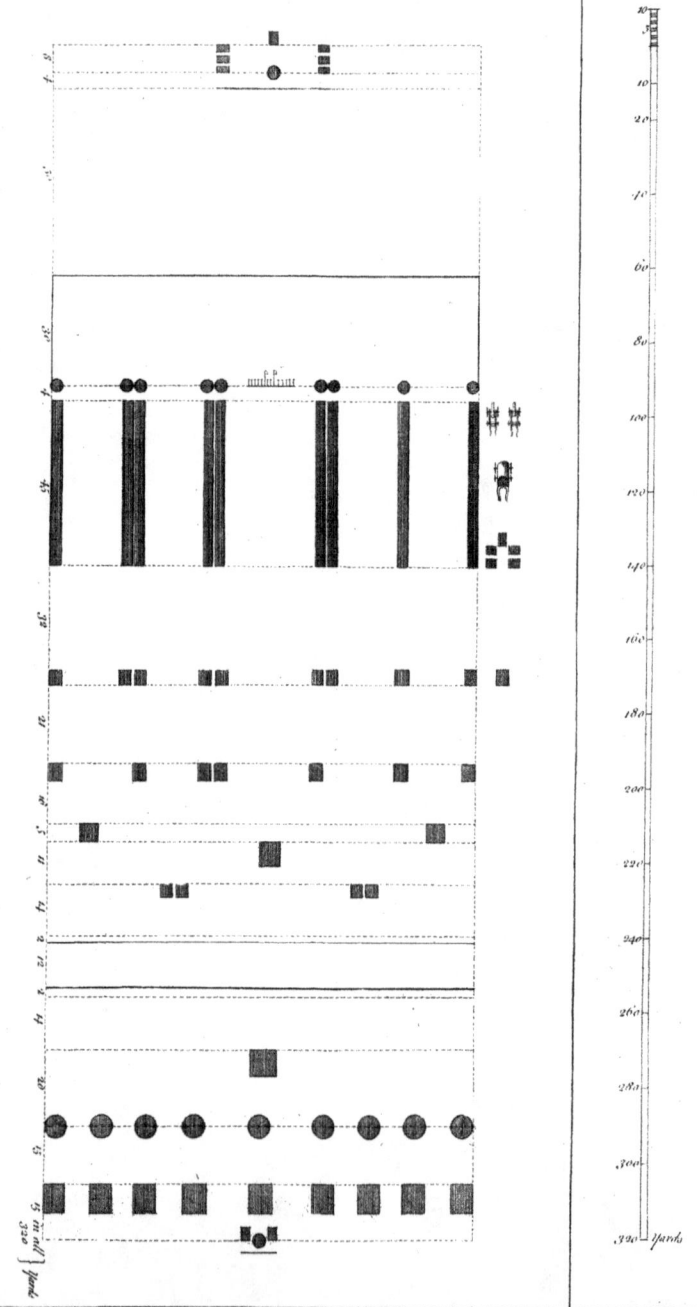

Plan of the Encampment of a Battalion of Foot of 9 Companies each consisting of 100 private Men.

Chap. XVII.   *Military Discipline.*   291

the several messes to boil the pot every day, for the reasons already mentioned in Article III. Chapter XIII.

I shall, in the following article, proceed to the particulars for the encamping of a Battalion of foot of nine Companies, of seventy-three men each, rank and file, with a plan of the same.

ARTICLE II.

	Yards.
For pitching three double rows of tents, at six yards each	18
For pitching three single rows, at three yards each	9
For the breadth of the grand street	25
For the breadth of four lesser streets, at seventeen yards each	68
Total front	120

	Yards.
From the side of Serjeants tent, to the center of the first gun	4
From the center of the first gun, to the center of the second	6
From the center of the second gun, to the left of the next Regiment	20
Total interval	30

Total front and interval 150

Depth 320 yards, divided as follows:

	Yards.
From the front pole of Officer's tent of quarter-guard, to the center of the bell of arms of ditto	8
From the center of the bell of arms, to the parade of the quarter-guard	4
From the parade of the quarter-guard, to the first line of the parade of the Battalion	50
From the first line of the parade, to the center of the bells of arms	30
From the center of the bells of arms, to the front of Serjeants tents	4

For

	Yards
For pitching fifteen tents, with their intervals, at three yards each	45
From the rear of the Battalion-tents, to the front of the Subalterns	32
From the front of the Subalterns, to the front of the Captains	21
From the front of the Captains, to the front of the Lieutenant-Colonel's and Major's	16
From the front of the Lieutenant-Colonel's, and Major's, to the front of the Colonel's	5
From the front of the Colonel's, to the front of the Staff-Officers	11
From the front of the Staff-Officers, to the front of the first row of batmen's tents	14
From the first row of batmen's tents, to the first row of pickets	2
From the first row of pickets, to the second	12
From the second row of pickets, to the second row of batmen's tents	2
From the second row of batmen's tents, to the front of the grand sutler	14
From the front of the grand sutler, to the center of the kitchens	20
From the center of the kitchens, to the front of ordinary sutlers	15
From the front of ordinary sutlers, to the center of the bell of arms, of the rear-guard	15
Total depth	320

The muzzles of the Battalion-guns, are in a line with the front of the Serjeants tents.

The rearmost of the gunners tents, are in a line with the rear of the Battalion-tents.

The Subaltern of the artillery, is in a line with the Subalterns of the Battalion.

The front-poles of the quarter-guard tents, are in a line with the front-poles of the center-companies, and in a line with the center of their bell of arms.

The bells of arms front the poles of the Serjeants tents.

The Colours and Drums are to be placed at the head of the grand or center street of the Battalion, and in a line with the bells of arms.

The two Companies on the right, and the Company on the left, form the three single rows; the other Companies form the double rows.

The Lieutenant-Colonel's, and Major's tents, front the center of the streets, on the right and left of the Battalion.

The Colonel's tent is in the line of the grand street, facing the Colours.

The Staff-Officers front the centers of the streets, on the right and left of the grand street.

The batmens tents front towards their horses.

The grand sutler is in the rear of the Colonel.

The inner diameter of the kitchens is sixteen feet, surrounded with a trench three feet broad, and the earth thrown inwards.—The two kitchens on the flanks, touch the outside line of the encampment. The center kitchen is in the center of the encampment, and distant sixteen yards and a half, from those on the right and left of it. The other kitchens are thirteen yards from center to center.

The front-poles of the ordinary sutlers tents, or huts, are in a line with the centers of the kitchens, allowing to each ordinary sutler, six yards in front, and eight in depth; enclosed with a trench one foot broad, and the earth thrown inwards.

The rear-guard fronts outwards.

The front poles are in a line with the center of their bell of arms, and distant from each other six yards.

The parade of the rear-guard is four yards from the bell of arms.

The houses of Office for the front line, must be advanced

vanced beyond the quarter-guard, at least fifty yards; and those for the rear line about the same distance in the rear of the petty sutlers, and butchers.

This plan being only calculated for the encampment of a Battalion, whose Companies are composed of no more than seventy-three rank and file, according to their usual strength in time of peace: I shall add another, to shew the method of encampment, made use of during the present war, while they remain augmented to a hundred each: and as the difference between them consists only in the quantity and division of the ground, it will be seen very plainly in the said plan, without any repetition of the preceding explanation, which, in all other respects, will answer for both.

There is a Serjeant of a Regiment, and a man of a Company, appointed to assist the Quarter-master, during the campaign, in marking out and keeping the camp clean; as also for the performing of all other things which appertain to their duty, such as the receiving of ammunition bread, or any other provision, which shall be distributed to their Regiments; all ammunition, working tools, carriages, clothes, and accoutrements; for which reason they do no other duty during the campaign, except on such where the Regiments mount entire.

The Serjeant is called the Quarter-Master's Serjeant, and the Soldiers, the camp-colour-men. Each camp-colour-man carries either a spade or a hatchet, which are delivered to them from the train.

When the army marches, the Quarter-Masters and the camp-colour-men are ordered before to take up the ground on which they are to encamp; and as soon as the Quarter-Master-General, or his deputies, have given them their ground, they are to mark out the encampment of their Regiments, and when that is done, they are to make the necessary-houses, and to get them finished, if possible, by the time the Regiments arrive, that the camp may be kept sweet and clean; for which reason the Sentries must have strict orders not to suffer any one to ease himself any where else; and when any Soldier,

Chap. XVII. *Military Discipline.*

dier, servant, or sutler, is found offending therein, the Commanding Officer should order him to be severely punished.

When the army continues any time in camp, new houses-of-office are to be made every sixth or seventh day, and the old ones carefully stopped up. The camp-colour-men are likewise to open a communication betwixt Regiment and Regiment, of a sufficient breadth for a grand division to march through in front, though they should remain but one night in a camp; but when that work proves too much for them to perform, the Majors must order them to be relieved by other men, and see the communication made.

The Quarter-Masters and their Serjeants are to see that the streets are swept clean every morning, and that the butchers and sutlers bury their garbage and filth every day, and that all dead horses are immediately buried, that the air may be kept from infection. They are likewise to examine the meat and drink which is sold by the butchers and sutlers, that they may not vend unwholesome provisions; and whenever they find any bad provisions, or any one attempting to sell it, they are immediately to seize both the provisions and the owner, and acquaint the Commanding-Officer with it, that he may give directions for their being tried by a Court-Martial, in order to their being severely punished for the same.

The Major of every Regiment is to inspect nicely into all these particulars, and not rely wholly on reports, but to see that they are punctually executed.

They are likewise to look into, and regulate the prices of all the provisions which are sold by the sutlers and butchers attending their Regiments, that the Soldiers may not be imposed upon. Neither are they to admit of any tents, huts or kitchens in the front, or any thing but the Quarter-guard, and the necessary houses.

The Quarter-Masters are to be very exact in entering regularly in their books, all the ammunition-bread or provisions, and all manner of stores which they shall receive for, and distribute to, their Regiments, that they may give an account of the several particulars when required.

ARTICLE III.

The *Reveille* is never beat the day the army marches, unless particularly ordered, but the *General* instead of it.

At the beating of the *General*, the Officers and Soldiers are to dress and prepare themselves for the march.

At the beating of the *Assembly* or *Troop*, they are to strike all their tents, pack up and load the baggage, and send it with a proper guard, to the place where the whole is appointed to assemble. After this, the quarter and standard guards, and the rear-guards, are to be drawn in, the Troops and Companies to draw up in their streets, and to be told off, that they may be ready to form into Squadron and Battalion at the next signal.

At the hour appointed for the army to march, the Drummers are to beat a *March* at the head of the line; and, as soon as they cease beating, the Squadrons and Battalions march out and form at the head of their encampment, complete their files, and tell off their Battalions by platoons, grand or sub-divisions, as it shall be ordered: and when the *March* is beat a second time on the right or left of the line, all the Squadrons and Battalions are to wheel towards the flank, where the *March* was beat, and begin the march as soon as wheeled.

The horse have different terms for the two first signals for the march of an army. The beating the *General* is called by the horse, sounding *to boot and saddle*; and the *Assembly* is, sounding *to horse:* however, in the general orders given out to the army, they are called by the terms which are used by the foot.

The usual time for the Regimental Quarter-Masters, the camp-colour-men, and the escort, to assemble, is, at the beating of the *General*.

When the army is to march towards the right, they then parade at the head of the right wing of horse; and, if they march to the left, they are to parade at the head of the left wing; but when the army is to march directly

towards

towards the front, the camp-colour-men, &c. parade then at the head of the firſt line of foot. Theſe are the general rules laid down for aſſembling of the camp-colour-men, and as ſoon as they are paraded, they are to march with the Quarter-Maſter-General to the place where the army is to encamp.

ARTICLE IV.

A little before the opening of the campaign, it is the duty of the Quarter-Maſter-General to draw out on paper the encampment of the army; in the doing of which he is to have a due regard to the ſeniority of the ſeveral corps; as alſo in placing the General Officers to their commands according to their rank: after which he is to preſent it to the General in chief for his approbation.

This plan, or draught of the encampment, is likewiſe called the line of battle; ſince the troops always encamp in the ſame order in which they draw up in the line, that, if the enemy, by a ſudden march, ſhould endeavour to ſurpriſe you in your camp, you may be ready to enter upon action as ſoon as you are formed at the head of your encampment: and though the General may think it neceſſary to alter the diſpoſition of his troops when he is going to attack the enemy, yet the encampment of the army is not changed on that account, but remains as at firſt fixed, unleſs other reaſons induce him to it.

In forming the encampment, the troops are divided into Brigades, and the Brigades into two lines, which are diſtinguiſhed by the firſt and ſecond, or front and rear lines.

A Brigade of foot generally conſiſts of four Battalions, and that of horſe or dragoons of ſix Squadrons; but, as the term of Battalion and Squadron is frequently uſed, though the number each conſiſts of is not mentioned, it is to be underſtood, that, in the general way of ſpeaking, a Battalion of foot is computed at ſix hundred men in rank and file; and a Squadron of horſe and dragoons at one hundred and fifty.

The

The method of forming the foot into Brigades, is as follows: The several Battalions are divided, according to seniority, into four equal parts or divisions.

The first part is to consist of the eldest Battalions; the second part of the next eldest; the third part of those next to the second; and the fourth part of the youngest Battalions.

The Battalions being thus divided into four classes, the first Brigade is composed of the eldest Battalion of each class; the second Brigade of the second Battalion of each class; the third Brigade of the third Battalion of each; and so on in this manner till the whole are formed into Brigades; by which method, there will be a Battalion of every class in each Brigade, and thereby intermix the old and young Battalions: for as entire Brigades are frequently detached, unless they are mixed in this manner, one composed of four young Battalions might be commanded on an affair of importance, and, for want of experience, fail of success; but by intermixing the experinced and unexperienced Battalions together, that danger is in a great measure avoided; which, in my opinion, shews the method not only right, but necessary.

The Battalions draw up in Brigade thus: The eldest Battalion is placed on the right of the Brigade, the second Battalion on the left of it, and the two youngest in the center, the third Battalion being on the right of the fourth. This rule, of placing the eldest Battalion on the right of the Brigade, is only observed by the Brigades which are posted in the right wing; but those in the left draw up the reverse, the eldest Battalion being posted on the left of the Brigade, and the second Battalion on the right of it, and so from left to right.

When the Brigades are formed, they are divided into two lines, as follows:

The first and second Brigades are posted on the flanks of the front line: and the third and fourth Brigades on the flanks of the rear line.

The

The fifth and sixth Brigades are placed in the front line, on the infide of the firft and fecond; and the feventh and eighth Brigades are placed in the rear line, on the infide of the third and fourth; and fo on in this manner till they are all formed in both lines, the youngeft Brigades drawing up in the center: for as the flanks of the lines are more liable to the attacks of an enemy, than the center, by their lying open, they are efteemed the pofts of honour, and therefore belong to the eldeft Brigades; but as the front line is more expofed than the rear, fince it begins the attack, while the other only fuftains it, the left flank of the front line, is undoubtedly, the fecond poft of honour, and therefore belongs to the fecond Brigade; fo that the right flank of the rear line can only be looked upon as the third poft of honour, and the left flank of the faid line, as the fourth.

This is the method when the troops which compofe the army belong to one Prince; but as the army in *Flanders* confifted of troops of feveral nations, every nation had a diftinct poft in the line; fo that the firft or eldeft nation had all their troops on the right; the fecond nation had all theirs on the left; the third had theirs on the left of the firft; the fourth on the right of the fecond, and the fifth, (if they confifted of fo many nations) had all theirs in the center. And though this may feem, at firft view, contrary to the foregoing rule, yet, by looking on every nation only as a Brigade (which muft be done in this cafe) it will be found, in every refpect, conformable to it.

The troops of each nation are generally divided in both lines, that thofe in the firft may be fuftained by their own troops; as alfo, that each nation may fhare equally of the danger; for as the front line is more expofed, in battles, that the rear, the placing the troops of any one nation entirely in the front line is never done, but on extraordinary occafions; it being reafonable to conclude, that their lofs will be greater than thofe in the rear line, whenever they engage.

The

The first nation posts their eldest Brigade on the right of the front line, and their second Brigade on the right of the rear line. Their third and fourth Brigades are placed on the left of the first and second Brigades, and so on by seniority till the two youngest Brigades are drawn up on the left of their own troops in both lines, the youngest posts being those which lie nearest the center.

The second nation draws up their two eldest Brigades on the left flank of both lines, and their two next Brigades on the right of the first and second Brigades, the left being with them the post of honour; so that their youngest Brigades close the right of their own troops.

The other nations observe the same rule, according as they are posted in the right or left wing.

Though the horse and dragoons now roll with one another upon every command, and go all under the denomination of cavalry, yet in the line of battle they are kept in distinct bodies, and placed in separate Brigades; it being a rule never to mix the horse and dragoons in the same Brigade.

The Regiments of horse are formed into Brigades in the same manner as the Battalions of foot; and though a Regiment of horse consists of several Squadrons, yet they are kept together in the same Brigade, and never divided, unless a Regiment should contain more Squadrons than a Brigade is generally composed of, as most of the *Imperial* Regiments do.

The same rule is observed by the Regiments of dragoons, in forming them into Brigades, as the horse are.

The horse are divided and encamped on the flanks of the foot on both lines, and posted according to seniority of Brigades, or Nations, on the right and left, as above directed for the foot.

The dragoons are divided and encamped on the flanks of the horse of both lines, and posted according to seniority, as the horse are; so that the dragoons are placed
on

on the extremity of the lines, and have thereby the post of the horse; but this is not given them by way of preeminence, but conveniency; for, as the dragoons, by their first institution, were only mounted upon little light horses, and designed for expeditious foot, they were therefore posted on the flanks of the army, that they might be ready to march on every occasion, such as convoys, covering parties, securing of passes or fords, or expeditions of the like nature, in order to save the horse for the most important acts of war, battles; in which the cavalry have so great a share, that they are generally either gained or lost by them; there being very few instances of the foot having gained a battle after the cavalry were beat.

These are the general rules for the forming an encampment, or line of battle.

In the encamping of an army, the first point which the Quarter-Master-General is to consider of, is the security of the camp, that it may not be liable to any sudden surprise, by leaving the flanks open and exposed to the enemy; for which end, it is usual to cover them with towns, villages, woods, morasses, or rivers, when such can be met with where the army is to encamp, that the enemy may not be able to approach the flanks without difficulty, or march a considerable body of men to attack them there.

The second consideration, is to have wood, water and forage near the camp, for the conveniency of the men and horses.

In the encamping an army, the front is to be always towards the enemy; but the troops which besiege a town, encamp with their front from the town, that they may be ready to draw out and oppose any succours which the enemy may endeavour to throw into it. Besides, as the guard of the trenches is always in proportion to the strength of the garrison, the rear of the camp is in no danger of *Sorties* from the town. This is the general rule; but when a town is besieged, where the besiegers have nothing to apprehend from without, by having
only

only the garrison to encounter, I presume they will then encamp with their front towards the town.

As the General Officers claim a right to have houses assigned them for their quarters, when towns or villages lie near the camp; one of the Quarter-Master-General's Deputies is always appointed to take up quarters for them; in the doing of which, he is to have a particular regard to the rank of each; and as soon as he has fixed upon the houses, he writes their names on the doors, and puts their respective Quarter-Masters in possession of them, every General Officer being to send one with him for that purpose.

The Train is generally encamped in the rear of the second line, and upon an eminence, that, if any accident should happen to the powder, the army may receive no damage by it.

Most nations have a Regiment belonging to the Train, composed of Gunners and Matrosses, and commanded by Artillery Officers; which Regiment never draws up in the line, or rolls with the army, but does only duty on the Train, and always encamps and marches with it, and at sieges they assist in erecting the batteries.

Though the Train attends on the army, yet it is a separate and distinct body, under the direction of their own Officers, and independant of every General in the army, but the Commander in chief, (always understanding by Commander in chief, the Officer commanding in chief a body of men with whom they shall be detached) whose orders they receive from his Adjutant-General, and not from the General Officers of the day, as the rest of the army does.

There is always an escort, which generally consists of horse, commanded with the camp-colour-men, to secure them from the enemy while they are marking out the ground, and till the army arrives.

If it is only a common escort, commanded by an Officer inferior to the Quarter-Master-General, the escort is then under the direction of the Quarter-Master-General,

Chap. XVII. *Military Discipline.* 303

neral, and posted by him as he shall think proper: but as considerable bodies of troops, commanded by General Officers, are detached with the camp-colour-men when they apprehend any opposition from the enemy, the General who commands the escort has the sole direction and posting of it.

When the Quarter-Master-General has taken a thorough view of the ground, and fixed the right and left of both lines, he generally leaves the rest to be performed by his Deputies, and goes with a party of horse to reconnoitre the country which lies towards the enemy, in order to see by what roads they can approach the camp, that he may acquaint the General with the several particulars as soon as he arrives, for his giving the necessary directions to the Generals of the day, to post a sufficient number of men on the grand-guard and out-posts, for the security of the camp, and the places where they are to be posted, which the Generals of the day are to see done.

The Quarter-Master-General is to reconnoitre the country to find out forage for the army, and to fix the places where they are to forage from time to time; as also to regulate the escorts, and the places to post them in, to secure the foragers: and when either of the wings, or the whole, is ordered to forage, the Quarter-Master General, or one of his Deputies, is always to go along with the Officer who commands the escort, in order to conduct him to his post, and to inform him with the situation of the country, that he may make a proper disposition of his men, both for keeping the foragers within due bounds, and to prevent their being fallen upon by the enemy.

The Quarter-Master-General is to provide guides to conduct the lines, artillery and baggage, when the army marches; as also for the foragers, and all detachments and out-posts that may require them.

In great armies there is always a company of guides established for that purpose; which company is under the care and direction of the Quarter-Master-General.

CHAP.

CHAP. XVIII.

Duty of the Trooops at a siege.

IN this Chapter I shall only treat of those things which relate to the workmen and the guard of the trenches, with some other particulars necessary to be known by those Officers who have not been on such important commands, in order to give them an idea of their duty.

As the method of carrying on the approaches is the duty of the Engineers, whoever would be informed of those particulars, must consult the works of Mr. *Vauban*, Mr. *Coborne*, and others who have excelled in that art.

ARTICLE I.

The foot are not only commanded as a guard to the trenches, but are likewise employed in the raising of the works, and the making of the several materials required, such as gabions, hurdles, sauciffons, fascines, and pickets: for the doing of which the men are regularly paid every day, or every two days at farthest, according to the following prices:

	l.	*s.*	*d.*
Each workman in the trenches had *per* night	0	0	8
Those who were employed on the batteries, had each in the twenty-four hours	0	1	4
The volunteers who were retained during the siege, to fill up the ditch of the town, laying a bridge over it, or such dangerous enterprises, had each every twenty-four hours, whether employed or not	0	2	6
Wool-pack volunteers, when employed had each	0	5	0

For

Chap. XVIII.   *Military Discipline.*   305

	l.	*s.*	*d.*
For the planting of a gabion, and filling it with earth, when the approaches were carried on by demi-sap	0	0	6
For the making a fascine and picket	0	0	3
For a sauciffon	0	0	6½
For a hurdle	0	0	8
For a gabion	0	1	4

Besides the charges abovementioned, there are a great many more which cannot be regularly stated; such as Miners; others employed in sinking of wells or pits to find out the mines of the enemy, for which they are sometimes paid half a pistole an hour; others hired to reconnoitre the breach, and fathom the ditch, if it is wet. When some desperate attack is to be made on a little out-work where a few men are only required, they generally do it by such who will go voluntarily, offering a reward to each man. Those who work in the sap have likewise extraordinary wages; with several others of the like nature, of which this nation is very little acquainted, because the *Dutch* were at the whole expences of the sieges during the late war; and as they were exceeding chargeable, the towns, when taken, were delivered into their hands in order to reimburse them; for which reason, the Field-Deputies of the *States* appointed a treasurer at every siege for the paying of the workmen, *&c.* from whom the Majors of Brigade received the money for those of their own nation, and gave it to the Majors of the Regiments, that they might clear their men.

A copy of the daily orders (in which is inserted the number of workmen of all kinds, as also the number of fascines, *&c.* to be made by every Regiment) was delivered to the treasurer, that he might know what was to be paid; which orders, together with the receipts of the Majors of Brigade, were to him sufficient vouchers for the payment of the money. These payments were made every day, or every two days at farthest, without which the works would go but slowly on: for though the men so employed run a great deal of danger, as well as undergo a great deal of fatigue, yet the desire of getting money does not only soften the labour, but makes the danger also appear less: but unless the men are punctually paid, it will be impossible for the Officers to keep them to their duty.

The men who are employed in the making the trenches, begin to work as soon as it is dark, and leave off at break of day; but when a work is not finished, and that the going on with it in the day-time may be done without exposing the men to too much danger, fresh detachments are ordered on when the others leave off, and go off at night when the others come on; so that they are only to work twelve hours, or the length of the night or the day. It is the same with those who work on the batteries; but as the cannon of the besieged fire constantly at the batteries, in order to dismount the guns planted on them, and thereby often damage both the parapet and platforms, they are therefore obliged to keep a sufficient number of workmen in a constant readiness to repair them, at whatever time it shall happen; which men are generally continued on that command twenty-four hours, and are therefore paid double wages, which is one shilling and four-pence each.

The working parties are always done by detachment, every Battalion at the siege, except those which mount the guard of the trenches, give an equal number of men to the works, with the same number of Officers and Serjeants to command them, as is usually ordered for other duties. Drummers are never commanded with these detachments.

The workmen march without arms, and carry only such tools as are proper for their works on which they are to be employed, such as pick-axes and spades; and the Officers who command them have only their swords and sashes; and notwithstanding these commands are both painful and dangerous, yet they do not pass for duties of honour, but only those of fatigue; and therefore

fore begin with the youngeft, as the others do with the eldeft. But before we proceed further, it will be proper to mention the neceffary preparation of materials, which are to be made by the foot, before the trenches are opened.

As foon as the town is invefted, bills and hatchets are delivered, by the train, to the feveral Regiments for the making of fafcines, &c. on the receiving of which, the Quarter-Mafters are to give receipts, that thofe which are not returned to the train at the end of the fiege, may be charged to the Regiments which do not.

Orders are then given for each Regiment to make fuch a number of fafcines, picquets, fauciffons, gabions and hurdles, in which the proper dimenfions of each is particularly fpecified, for the information of thofe who are to make them; the common proportions of which I fhall here infert, and are as follows:

A fafcine, is only a faggot made of the green-branches of trees, about fix feet long, with four or five bindings, and of the thicknefs of a common faggot. They are not only ufed in the parapet of the trenches and batteries, but likewife in the filling up of ditches for the paffing of them.

Pickets, are ftakes about four feet long, and made fharp at one end, by which the fafcines are faftened to the ground.

A fauciffon, is a fafcine of about fixteen feet long, made of very fmall branches, and no thicker than an ordinary fafcine, having the bindings within fixteen or eighteen inches of one another. They are ufed in the platforms of batteries, and for making of blinds, when any of the works are enfiladed.

Hurdles are fo well known, from their being conftantly ufed in fheep-folds, that there is no occafion for an explanation. When the trenches are very wet or dirty, they are laid at the bottom of them, for the conveniency of walking. They are alfo ufed for the paffing of moraffes.

Gabions are round wattled baskets, open both at the top and bottom. They are of different sizes; but those which are generally made use of, are about four feet diameter, and five or six feet high. They place them frequently on the the platforms of the batteries, and fill them with earth, which makes the parapet, or, as is called in the terms of fortification, the merlons of the battery. The merlons, are only the pieces of the parapet or wall which rise up between the embrasures to secure the gunners from the fire of the besieged. Gabions are always used when the approaches are carried on by demi-sap; or, when they cannot sink a trench, by meeting with wet or marshy ground, they make use of Gabions to carry on the approaches above-ground. Woolpacks and sand-bags are also proper in such places, as well as for the making of lodgments.

When the orders are given for making fascines, &c. every Regiment is to send out proper detachments to the adjacent woods, or places appointed for that purpose; and when the men have made the number ordered, the Officers are to return with their detachments to their Regiments; after this, fresh detachments are to be sent every day till the whole number directed it made; and so on during the siege, for whatever number shall be wanting.

The Officers who command these detachments, are to give in a list of the men to their Majors, with the number of fascines, saucissons, hurdles, and gabions, placed opposite to the names of the men who made them, that they may be paid for the number they have made, as soon as he receives the money. The same rule must be observed, in giving in a list to the Major, by the Officers who command the workmen in the trenches, &c. or by the Adjutant when they are detached from the head of the Regiment: by doing of which, no dispute can happen in paying them, nor give them the least room to think they are wronged; a circumstance of no small importance to the service, since a contrary proceeding is often attended either with mutiny or desertion; and

therefore

therefore every cause that can incite them to it, ought to be carefully avoided.

Detachments of horse and dragoons are ordered at the same time with the foot, to carry the fascines from the places where they are made to the general magazine of the trenches, which is commonly at or near the place where they intend to open them.

These detachments are likewise commanded by Officers; but the Troopers and Dragoons carry neither swords, carbines nor pistols. Every one carries a fascine and a picket, (which they lay before them on the pummel of their saddles) and march in file to the place where they are ordered to lay them down; after which they return for more, and so on, till they bring the number ordered. But when this proves too fatiguing to the cavalry, the peasants are summoned in with their waggons, as also those belonging to the train and bread, in order to be employed in carrying them; and, on extraordinary occasions, all the waggons belonging to the sutlers are likewise sent, and the General Officers are desired to send theirs.

As the gabions and saucissons cannot be carried on horses, waggons from the train are always sent for them.

When the woods or places where the fascines are made lie near the camp, the men who make them are frequently ordered to bring them to the head of their own Regiments; from whence they are carried, as abovementioned, to the general magazine of the trenches; to which place the working tools are likewise to be sent from the train, where commissaries or storekeepers are constantly to attend for delivering of them to the workmen, and receiving them back when they leave off: for which reason, the Officers who command the workmen are, when they leave off work, to march their detachments to the said magazine, and deliver the working tools to the storekeepers.

A guard of foot is always placed over the said magazine, to prevent any of the tools, or other materials, being taken from thence, but by the direction of the proper Officers.

ARTICLE II.

When the trenches are to be opened (which is always made a very great secret, that the besieged may not know the certain time of doing it) a sufficient number of Battalions and Squadrons, with General Officers in proportion to command them, are ordered for the covering party, to secure the workmen from the *Sorties* of the besieged.

A little after sun-set, the said Battalions and Squadrons draw out at the head of their encampment, and march from thence so as to arrive at the general rendezvous just as it grows dark, where they are joined by the General Officers who are to command them; as also by the Engineers who are to trace out the works.

The rendezvous so appointed, is generally at or near the place where the trenches are to be opened, and is therefore called afterwards, the parade of the trenches.

When the covering party is arrived at the parade, the chief Engineer who has the conducting of the siege, acquaints the General who commands the troops with the place where he intends to open the trenches, and how far he proposes to carry them on, with the situation of the ground betwixt that and the town, that he may post his troops accordingly: for as the Engineers are obliged to reconnoitre the ground thoroughly day after day before they can determine where to make the attack, they can therefore inform the General of the places where the Regiments may be posted to the best advantage, both for their own security as well as that of the workmen.

The men are to keep a profound silence both in the marching to take possession of their posts, and, during the time they continue there, to prevent their being discovered by the enemy; and, as soon as the foot have placed a sufficient number of Sentries, to give notice when a *Sortie* is made from the town, they are to lie flat on the ground with their arms in their hands, unless their situation is such as to cover them from the fire of the enemy, by being posted behind some rising ground,

hedge,

Chap. XVIII. *Military Discipline.* 311

hedge, ditch, wall, or old ruins; but when those are not to be met with, their lying flat on the ground will be a means of saving a great many of them, since they cannot be easily discovered in that position; and if they are not, only the dropping or spent ball will not touch them; but supposing they knew where they were posted, they will receive much less damage by lying down, than if they stood up, as every man's reason must immediately suggest to him; and therefore I shall not trouble the reader with further arguments to prove it. The horse cannot pursue this method; for they must continue mounted all night, that, when a *Sortie* is made, one part may be ready to sustain the foot, while the other endeavours to get between the enemy and the town, to cut off their retreat.

The workmen are ordered at the same time with the covering party; for which service every Battalion then off duty, is to furnish an equal number, such as an hundred, a hundred and fifty, or two hundred men each, according to the works which they intend to throw up that night; which detachments are to be on the parade of the trenches, just as the covering-party marches from thence to their posts, and to draw up (either six or three deep) according to seniority of Regiments, which the Major of Brigade of the day is to see done, and to examine their numbers, to know if they have complied with the orders, in sending the proper complement.

As soon as the workmen are paraded, pick-axes and spades are delivered to them; in the doing of which a regard is had to the nature of the ground, that if it is hard or rocky they may give a greater proportion of pick-axes; but when of a soft and loose mold, the number of spades exceeds the other. In the distributing of the tools to the men, the Officers should take care to intermix them in such a manner, that as fast as one man loosens the earth with a pick-ax, the one next him may have a spade to throw it up with; and as this rule must be observed in the drawing up of the men to work, the

inter-

intermixing of the tools on the parade will save the trouble and time of doing it afterwards; after this, a fascine and a picket is delivered to each man.

As soon as the covering-party is posted, the Engineers trace out the approaches; but as the doing of it with common lines would be of little use in the night, from their not being easily seen, they trace them out with straw lines, that is, ropes made of straw. After this, they return to the parade, and acquaint the Major of Brigade of the day what number of men with tools they must have to the several parts, and what number without tools, for the carrying of fascines, &c. and when the disposition of the workmen is made, the Engineers conduct them to the several parts traced out (each man carrying a fascine and a picket with him, besides his working tool) where the Officers are to draw up their detachments in a single rank behind the line and facing it, leaving an interval of three feet betwixt each man.

Note. The meaning of the expression, *behind the line* or *trenches*, is the side from the town, or next the camp; and when the word *before* is used, is understood the side next the town.

When the men are drawn up, as above directed, they are to place the fascines along the straw line, and fasten them to the ground with the pickets. After this is done, the men with the pick-axes are to dig holes behind them about four feet from the line on which the fascines were laid; and as fast as they loosen the earth, the men with the spades are to throw it on the fascines; and as soon as the holes are about a foot, or a foot and a half deep, the men with the pick-axes are to stand in them, and dig the ground which lies between the holes, till they are all opened into one another; which being done, forms a trench or ditch running parallel to the line on which the fascines were placed; and the earth which comes out of the said trench being thrown on the fascines, raises the parapet or breast-work betwixt them and the town; but as the men will be vastly exposed, if they are discovered, till the depth of the trench and

the

the height of the pararet is sufficient to cover them from the fire of the besieged, the Officers must therefore oblige them to keep a profound silence, and apply themselves thoroughly to the work, till that is done; after which, they may allow them to go on more moderately in the finishing of it, though without suffering them to be idle, which, after they are covered from the fire of the enemy, is generally the case, unless they are well looked to. It is therefore the duty of the Officers on these commands, to walk constantly from place to place, in order to view every part of the work on which their detachments are employed, since on their diligence, that of their men will depend.

The men who are appointed to carry fascines only, are to bring them from the general magazine to the places where the others are at work; in doing of which, the Officers who command them, are to take care that they are diligent, lest those who are at work should be forced to stop for want of fascines to raise the parapet; the consequence of which may occasion you the loss of a great many men, by being longer exposed to the fire of the town than they would be when they are duly supplied.

As soon as the fascines, which were laid down first, are covered with earth; another row or line of fascines must be placed on the top of them; and fastened down with pickets; and when these are covered with earth, a third row of fascines must be placed on them; and so on in this manner, by intermixing of earth and fascines, till the parapet is raised to its proper height; which is breast-high from the foundation; which, if fascines are not wanting, and that the ground is not excessively hard or rocky, may be done in a very short time, at least so high as to cover them from the fire.

The reason of their using fascines in the parapet, is not only for the raising of it quicker, but likewise to strengthen and support the earth, which, being loose, could not stand without them. Besides, as the inside of the parapet must only have a little slope, like the escarpe of a

ram-

rampart, fascines are absolutely necessary, since the earth would be continually falling down without them.

From the top of the parapet, the earth must run with a gentle slope, like those commonly made in gardens, towards the town. The reason for this is, that when the enemy come to attack the trenches, they may find nothing to cover them from your fire; whereas if the parapet had no greater a slope on the outside than on the inside, it would serve as a breast-work to the enemy when they come to attack you.

The banquet, or foot-bank, is the space of ground which is left standing betwixt the parapet and the brink of the trench, and should be at least three feet broad after the parapet is finished. It serves as a basis to the parapet, and for the Sentries to stand on, to discover what passes betwixt the trenches and the town.

The trench should not be above three feet deep in the solid ground, or rather less, if you can raise the parapet high enough without going so low, since the sinking of a deep trench seldom fails of making it exceeding dirty. It should be at least four feet broad at the bottom, that the Soldiers may march in it two a-breast, or pass by one another with ease.

These are the common proportions of the several parts, the whole of which is called the Trenches: and though it is not only the duty of the Engineers to instruct the Officers, that they may direct their men in the making of the works; but likewise to visit them from time to time, to see that each part has its true proportion; yet as those gentlemen are liable to accidents, from the danger they are often exposed to, that may render them incapable of performing their duty, by which the works may be retarded, or ill executed, unless the Officers, from their own experience, can supply the defect; I thought the inserting the above particulars would be of use to young Officers, by giving them some notion what trenches are, and in what manner they are made, that, when they shall be ordered on those commands, they may not be entirely at a loss how they are to proceed,

in

Chap. XVIII. *Military Discipline.* 315

in cafe they fhould fail of the neceffary directions and affiftance of the Engineers, which proves too often the cafe at moft fieges; nor is it to be much wondered at, for if the Engineers do their duty, they are fo often difabled, that their want muft be fupplied by the diligence and fkill of the Officers.

At break of day the workmen leave off, and the Officers march their detachments back to the camp, making a report firft to the General who commands in the trenches what number of their men are killed and wounded, and the fame to their own Colonels as foon as they difmifs their men.

At the fame time that the workmen leave off, the Battalions, which covered them, take poffeffion of the trenches, at leaft fuch a number as fhall be thought fufficient, which are always the eldeft, and the reft return to the camp. The Squadrons are likewife drawn off, leaving a proper detachment, if thought neceffary.

When the trenches are opened, the workmen are to be on the parade of the trenches always at fun-fet. The Major of Brigade of the day is to be there at the fame time, to look into their numbers, and to detach them from thence to the feveral works, as the Engineers fhall direct; to which they are conducted by old experienced Serjeants, who are employed by the Major of the trenches to look after the works, and to affift him in the performing of the feveral parts of his duty, and are therefore called, the Adjutants of the trenches, for which they have extraordinary pay.

The common method of detaching the workmen from the parade of the trenches, is, by beginning on the right, and fo on to the left, till the whole are difpofed of; but as this method proves very prejudicial to fome Regiments, by having their entire detachment fent to thofe works which are vaftly expofed to the fire of the befieged, and thereby lofe a confiderable number of their men, while others receive little or no damage; I fhould therefore think it highly reafonable, when this is the cafe, that an equal number of men fhould be taken from the

detach-

detachment of each Regiment, and sent to the works which are most exposed, by which means the loss will fall more equally on the whole: otherwise the Regiments which are drawn up on the right will be great sufferers, particularly when the approaches are carried near the *Glacis*, since those who are first detached, are generally employed on those works, and the others on things of less moment, such as the finishing of works already begun, or the carrying of fascines, &c.

I own, that the detaching of the workmen according to the method proposed, will create the Major of Brigade a little more trouble; but surely that cannot be given as a good reason for its not being done: for where the lives of men are in question, impartial justice should be done them, whatever pains it may cost.

The posting of Battalions betwixt the workmen and the town, is only done on the first breaking of ground, they being afterwards to remain constantly in the trenches; however, they sometimes make detachments from the Regiments, of Lieutenants and Captains commands, to support the workmen, and post them at the head of the trenches, that they may be ready to march out and attack those who sally from the town, in order to put a stop to their progress till the Battalions can be brought up, to facilitate the retreat of the workmen, and to prevent their being pursued.

When a *Sortie* is made, which obliges the workmen to quit their posts (which a very small matter will do) the Officers who command them, are to endeavour all they can to prevent their dispersing: for which end, every Officer is to draw his detachment into a body, and retire with them either into the next trench, or at a proper distance from the works; where they are to remain till the enemy are repulsed, and then return with them to their former stations, in order to finish what they had begun. But as orders, on these occasions, are not of sufficient force to oblige the men to keep with their Officers, or return with them to their work, after the enemy retire, unless proper punishments are annexed to the

breach

breach of them, and, when found out, strictly executed, and when neglected, proves a great detriment to the service, by prolonging of the siege: to prevent this, some other expedient, than what has been hitherto found out, should be thought of to keep the workmen within moderate bounds: and though this is a difficult task, yet, I believe, it may, in some degree, be effected, by inflicting some corporal punishment on those who shall do it (unless they are wounded) and stopping of their working money, and dividing it amongst those who remain with their Officers, as a reward for their complying with their duty, and likewise to induce them to detect those who shall quit their commands; for without the money is divided in this manner, they will not detect one another, but rather endeavour to conceal it from their Officers. As this method was never practised, that I know of, I therefore only offer it as a notion of my own, in order to put those who shall command, on these occasions, upon finding out a proper remedy against an evil which is constantly complained of, though still suffered to go on.

ARTICLE III.

The guard of the trenches is never formed by detachment, but is always composed of entire Battalions; the number of which must depend on the strength of the besieged, it being necessary to have a sufficient force to repulse any *Sortie* they shall be able to make, either for the levelling of the trenches already made, to interrupt the progress of those carrying on, or nailing up of the Cannon on the batteries.

As the mounting of the guard of the trenches is a duty of honour, it always begins with the eldest Regiments.

The guard of the trenches is only a duty of twenty-four hours; but the time of relieving it is not fixed, since it is sometimes done in the morning, and sometimes in the evening; though formerly the relief was always made

made in the dusk of the evening, to avoid, as they said, the danger which they must run from the fire of the besieged, if they made it when it was light; but as they did not observe that rule at the sieges in *Flanders*, we must conclude, that doing it always in the dusk of the evening proceeded rather from custom than reason: for when the relief can be made with safety in the day-time, it is much properer than when it is dark, since they cannot then make so good a judgment of their posts, or know by what communication they can march to sustain one another, in case of a *Sortie*, as when they mount in the morning, or some time before it is dark. It is true, when they mounted in the dusk of the evening, the Majors of those Regiments took a view of the trenches in the morning, to inform themselves thoroughly of all the particulars relating to their posts; as also those belonging to the other Regiments, that, by knowing the situation of the whole, they might the more readily march to sustain one another, on the first order they should receive from the General who commands in the trenches: after this they returned to the camp, and made a report of the same to their respective Colonels.

After the trenches are opened, the Regiments which mount, are always named the day before in public orders, that they may be prepared for it: neither do they furnish any men to the ordinary or extraordinary guards (their own quarter-guards excepted) or working parties, after such orders are given, till they are relieved, that they may march into the trenches as strong as possible, leaving no more men on their quarter-guards than what are absolutely necessary for the security of their tents and baggage.

General Officers are appointed at the same time to command the troops in the trenches, who mount and dismount with them. The number of General Officers are in proportion to the number of Battalions which mount; and when the attack is very considerable there is then a Lieutenant-General, and a Major-General ordered for the command, with whom a Major of Brigade tre-frequently mounts, for the receiving and delivering of the General's orders to the troops in the trenches, otherwise the eldest Major commonly performs that duty, unless the General appoints another.

When there are more attacks than one carried on at the same time, a proper number of General-Officers are appointed for the command of each.

The Regiments which mount the trenches, are to march from their own encampment, so as to be on the parade an hour before the time appointed for the relief, where they are joined by the Generals who are to command them.

When the parade of the trenches is not naturally covered from the Cannon of the besieged, by having a rising ground before it (which they always pitch upon for the opening of the trenches, when such a place can be found near the attack) a large epaulement is then thrown up to secure those who come to relieve, and that they may march from thence into the trenches without being exposed to the fire of the town.

The Regiments are to draw up by seniority on the parade of the trenches; and as soon as they are all formed, whatever detachments are to be then made from them, either for guards to the batteries, magazines, or particular advanced posts, are to be immediately drawn out, with proper Officers to command them, and formed by the Major of Brigade who mounts the trenches, or the Officer who is appointed to do that duty; after which the Regiments are to be told off into platoons, and the Officers divided equally to them, with which they are constantly to remain till they are relieved.

The Regiments in the trenches are to send each a Serjeant to the parade, to conduct those who come to relieve them to their several posts; the trenches being sometimes of so great an extent, and so intricate, that without this precaution, it will be difficult for them to find their way, particularly when the approaches are advanced up to the *Glacis*.

When

When the time of relieving is come, the General orders the Regiments to march and relieve the guard of the trenches.

They march in by seniority, the eldest relieving that which is posted at the head of the trenches, or most advanced parallel. The rest are to follow the same rule, the posts of honour being those which lie nearest the town.

When they come to the *queuë*, or tail of the trenches, the Grenadiers, led by their Officers, are to rank off singly from the right, or march two a-breast, if the trench is wide enough to allow of it; at the bottom of which they are to march, to void being exposed to the fire of the besieged: for though they are not to be afraid of the fire, yet the Soldiers are never suffered to expose themselves, but when the service requires it; and then they are not to decline it, though they should be sure of meeting with certain death.

The Battalions are to march into the trenches in the same manner, the Officers keeping with their respective platoons: and when they come opposite to the Regiments which they are to relieve, they are to halt, and face them. After this, the Sentries posted on the banquet, to discover what passes betwixt the trenches and the town, are relieved; and the Colonels who are to be relieved, acquaint those who come to relieve them with all things relating to their posts, what additional Night-Sentries are necessary for their further security, &c. during which time, the Majors and Adjutants, accompanied by the others, take an exact view of the works where their Regiments are posted, that if any part of them are damaged or broke down, or that the parapet is not thick enough to resist the cannon of the besieged, they may apply to the General of the trenches for workmen to repair them, or obtain his leave for the doing it by a detachment of their own men, since none of those, who belong to the guard of the trenches, can quit their arms, or leave their posts, without his directions. They are likewise to reconnoitre the several communications lead-

ing

ing to the other Regiments, that if any should be attacked, they may know, when ordered, how to sustain them without loss of time. When the Majors have looked into all these particulars, (which ought by no means to be omitted, since several of their men may be lost for want of a due inspection into the works) they are to make a report of the same to their Colonels; after which, the Regiments that are relieved march out of the trenches, and return to the camp.

As soon as the old guard is marched off, those of the new are ordered to sit down on the banquet, holding their arms between their legs, which they are not to quit, or stir from their posts, but on occasions of necessity; and even then not without leave of their Officers: nor must they be suffered to sleep, that they may be always ready to oppose every attempt which the enemy shall make upon the works: and, on the first noise, or notice given of a *Sortie*, the Soldiers are to stand up in the trench; and, if the noise increases, or that the Sentries confirm the report, the Battalions are to form on the reverse of the trench (which is the side of the trench opposite to the banquet) and remain there till the General of the trenches shall send them such orders as he shall think proper.

The relief of the trenches is always made without beat of Drum, and with as little noise as possible, that the besieged may not know the exact parts you are in, by which means their fire can be given only at random; whereas, if the Drum was to beat, or a considerable noise made, they could direct it with more certainty, and thereby make your loss greater.

Though it was the constant practice formerly, for the Regiments to march into the trenches with flying Colours, and planting them on the top of the parapet, as soon as they had taken possession of their posts; yet, towards the latter end of the late war in *Flanders*, that ceremony was laid aside, and the Colours left in camp under the care of their own quarter-guard, or sent to the Regiment which encamped next them; having found, by experience, that it did not only show plainly where the Regiments were posted, but proved likewise a temptation to the Gunners of the town to point their cannon at them, by which they lost a considerable number of their men, particularly the Sentries who were posted on the banquet: and, as the carrying them to the trenches was not in the least essential, but rather destructive to the service, by exposing their men without any real or probable advantage, nothing could be more just and reasonable than the quitting of that ceremony, since the *English* do not want such shows to animate or spur them on to their duty. The *French*, however, keep up the old custom of carrying their Colours with them, from a notion, perhaps, that it looks more daring.

Though it is reasonable to suppose, that the care which lies on the Generals who command in the trenches will oblige them to move frequently about the several posts, in order to keep the troops alert, and, by their presence, to animate the workmen who are carrying on the approaches; yet there is a fixed place where they are to be sent to on every occasion, which is generally at or near the Battalion, which is posted in the center, as being the most convenient in the sending of orders to, or the receiving of, reports from the whole, and where one of them (when there are several on that command) should always remain for that purpose.

Whatever intelligence the Colonels shall receive of the motions of the besieged, they are to send an account of it immediately, by an Officer, to the General of the trenches, that he may give the necessary directions to the whole.

When the besieged intend to make a vigorous sally for the levelling of your works, or the nailing of the cannon on the batteries, it is frequently preceded by some sham ones; their design in which, being to amuse, or draw off your attention so far, that when they make the real one, you may image it such as the former, and, by that means, neglect the necessary precautions for your defence; therefore an Officer must not suffer

himself

Chap. XVIII. *Military Discipline.* 323

himself to be imposed upon by that, or any other stratagem of the like nature. Besides, they may just act the contrary, by making the real one first; so that there is no knowing when they will make it, or judging of their future actions by their past; for though *Sorties* are generally undertaken in the night, yet there are instances of some being made at noon, which (from their not being expected at such a time) have proved more successful than the others. I only mention this to shew how necessary it is for the troops in the trenches to be always on their guard, and not to imagine themselves in a state of security from the enterprises of the besieged, while they are on that duty.

At some of the great sieges in *Flanders*, in the late war, besides the General Officers already named for the trenches, a General of the foot was appointed for the command of each attack, who had the care and direction of it during the whole siege; but as this is quite out of the common rule of the service, by its never having been practised before, that I know of, I shall not pretend to determine on its being right or wrong, or trouble the reader with a particular detail of their duty.

When the attacks were commanded by Generals of the foot in this manner, the attacks were called after their names; otherwise they were distinguished by the right, left, or center attack; or by the names of the Engineers who had the conducting of them; or by the names of the bastions where the breaches were made.

A little before the trenches are relieved, every Regiment in them is to send a return to the General, of what men have been killed or wounded during the time they have been on duty. The Officers who command the several detachments of workmen, are to do the same when they leave off work; as also the Artillery-Officers who command in the batteries; that he may acquaint the General who commands the siege, with the several particulars, when he makes his report to him, which is always done as soon as they are relieved.

When a *French* Battalion is to mount the guard of the trenches, it is always formed in picquets, of forty-eight, or fifty men each, inſtead of platoons.

Theſe picquets are compoſed of an equal number of men from each company, in the ſame manner as the ordinary picquets are formed.

The picquet, which is drawn up on the right of the Battalion, is compoſed of the Officers and Soldiers who are to go firſt on duty. The ſecond picquet, which is drawn up on the left of the firſt, conſiſts of thoſe who are the next on command. All the other picquets are ormed in the ſame manner, and are drawn up in Battaion according to their tour of duty, thoſe who are to go on laſt being on the left.

The reaſons which they give for it are theſe: when detachments are wanted in the trenches, no time is loſt in making them, they being formed, and ready to march on the firſt order. Beſides, ſhould an accident happen to any part of the Reigiment by a *Sortie*, the loſs will not fall on particular companies, but equally on the whole.

Theſe are the reaſons which they give for forming the Battalions that mount the trenches in this manner; and which, in my opinion, ſeem to carry a good deal of weight; but as no other troops but the *French* follow the ſame rule, I ſhall be ſilent on that head, leaving every one to judge of it as they ſhall think proper.

ARTICLE IV.

Beſides the foot which are ordered for a guard to the trenches, there is alſo a body of horſe commanded to each attack, who are relieved every twenty-four hours, as the foot are, and are under the direction of the General who commands in the trenches.

They are always poſted at or near the *Queuë* of the trenches, ſo as not to be expoſed to the cannon of the beſieged.

As ſoon as they mount, the Officer who commands them ſends a Cornet on foot into the trenches to attend

on

on the General, that when he has any orders for the horse, he may send them by the Cornet.

The number of horse which mount, are more or less, according as it shall be judged necessary, they being designed, when the enemy attack the trenches, to cut off their retreat to the town; or, if the *Sortie* is composed of horse and foot, you may have horse to oppose theirs; so that a regard must be had to the quality of the troops in garrison, and the nature of the ground which lies near the approaches; for if it is not plain and open, neither their horse nor yours can act, and therefore a small number will be sufficient: but when the garrison is strong in horse, and the ground proper for them to act in, a greater body of horse is required on the trench-guard. But as the *French*, during the late war, depended more on the strength and regularity of their fortifications than on a numerous garrison for the prolonging of a siege, they seldom threw more troops into a town than what were sufficient to man the works; so that they very rarely made a considerable sally, knowing that if an attempt of that kind should miscarry (which more frequently happens than not, if the troops in the trenches do their duty) the loss of the town would soon follow, by reducing the garrison too low to make a proper defence: and as this was the case at most of the sieges in *Flanders*, the horse-guard seldom consisted of above two hundred at each attack, (after the trenches were opened) but oftener much less, there being very few instances, if any, where the service, for which they mounted, was required; and when it is not, the mounting of a great number proves detrimental to the service, by putting them on unnecessary fatigue, since a great many horses are thereby entirely lost, and others reduced to so low a state, as to render them unfit for present service. However, neither this, nor any other consideration ought to prevail, or be allowed as an excuse for their not mounting a sufficient number on the trench-guard, whenever there is reason to apprehend they may be wanted. But notwithstandng the old custom of making great *Sorties*, for levelling the works, and nailing the cannon, was, in a manner, laid aside; yet the *French* Governors fell upon another method to retard the progress of the works without much danger to the garrison; which was, by making of frequent *Sorties* in the night, of an Officer and twenty men, or a Serjeant and twelve, who marched up to the workmen, cried out, *Tué! Tué!* with a loud voice, then fired amongst them, and immediately ran back into the covert-way; and, as they were only sent to disturb the workmen, they had generally the desired effect; for, upon these *Sorties* only, the workmen could not be kept to their duty by their Officers, but threw down the tools and dispersed; after which, few or none could be found again that night to go on with the works; so that without any danger to the garrison, the progress of the siege was as effectually retarded as it could be by great sallies, unless they could make themselves masters of the Batteries.

These small *Sorties* were sometimes made three or four times a night; and to encourage the men to undertake them, the Governors always gave a crown or ten shillings a man to those who would go voluntarily; and though they suffered for it now and then, by the troops in the trenches, yet that never-failing argument, money, procured them always a sufficient number of the most bold and intrepid to offer themselves; by which method the workmen were so often interrupted and alarmed, that some nights they did not finish one quarter of the works which were begun; and notwithstanding this was repeated night after night, yet I never heard of any expedient fallen upon to stop it; from whence I conclude it was found impracticable; for as it could only be done by posting of small guards, commanded by Serjeants, betwixt the workmen and the town, in order to intercept those who sally, yet as it was not done, I suppose it proceeded from the too great danger which those guards must be exposed to from the fire of the besieged; otherwise it was impossible that so clear a point could be overlooked.

There being frequent occasions at a siege to make small

small attacks to dislodge the enemy from their advanced posts, or little detached works, which interrupt the besiegers in the carrying on of their approaches; as often as attacks of this kind are necessary, they are always performed by the guard of the trenches; for which end, every Battalion then on duty must give their proportion of Grenadiers and others, according to the number which shall be ordered, that the loss may fall equally on the whole; and not by particular Regiments, on account of seniority, as was formerly practised, that custom being entirely laid aside, and making them by detachment instituted in their room, as a more just and reasonable way of proceeding; otherwise the oldest Regiments must be tore to pieces, while the young ones were safe in the trenches, looking on as spectators. Besides, unless this method had been altered, the *English* must have been constantly destroyed at the sieges in *Flanders*: for, as the guard of the trenches was composed of the troops of different nations, of whom the *English* took post, as belonging to the oldest crowned head, or eldest nation, as they called it, all those attacks must have been made by them; which sufficiently evinces the absurdity of the old method, and the equity of the new, in making every Battalion then on duty share equally of the danger.

On particular occasions, when the affair was very dangerous, and that a small number of men was only required, it was done by volunteers (that is, such who would go voluntarily) offering a reward to each man; but even in this case those who command Regiments should not suffer any more volunteers to go than their proportion; otherwise they may lose a considerable number of their bravest men, by the temptation of money, while the Officers, who act with more caution and prudence, save theirs, by not allowing them to offer themselves till they are called upon, in hopes that their places will be supplied by those of other Regiments; several instances of which could be easily given, were there an occasion.

The same rule should be observed in relation to those who are employed in the sap, demi-sap, carrying of wool-packs, and the half-crown volunteers, (as they are called) since every Regiment can supply their proportion of men duly qualified for such undertakings; but as some may have neither artificers nor miners, they are always taken where they can be found, without any regard to the said rule.

The number of half-crown volunteers are more or less, according to the business which may be required of them, which is to fill the ditch with fascines when the breach is made, in order to attack it; as also to lay bridges over it, for the same purpose: and though these men have nothing to do till towards the end of the siege, yet they are retained, and paid from the opening of the trenches, and are excused from all other duty whatever till the town is taken, or the siege raised: but they pay dearly for it whenever they are employed, it being hardly possible for them to escape.

My reason for inserting the above particulars, is principally designed to inform those, who have not been on such commands, of the proper method of proceeding, that when they shall be engaged on the like occasions they may know how to conduct themselves accordingly; without which, they may be easily caught by the specious pretences, or compliments of designing men, in putting the old puncto of seniority upon them, to the no small detriment of those under their care. The same may likewise happen in case of volunteers, by allowing a greater number than your proportion: for the hope of reward is so strongly implanted by nature, that it creates in mankind even a contempt of death when the prospect is in view, as was frequently seen by the surprising actions which were performed at the sieges in *Flanders*, by the giving of money; which, when duly regulated, is exceedingly proper, and proved of great service in taking the towns much sooner than they otherwise could have done; therefore it is not the method of giving money which is wrong, but the ill use that may be made of it, by imposing upon particular Regiments, unless the Officers who command them are aware of the bait.

ARTICLE

ARTICLE V.

The preparations which are generally made for an aſſault on a conſiderable outwork, or the body of the place, are as follows:

The number of troops which are commanded on theſe occaſions, muſt depend on the ſtrength of the place to be attacked, and the number of men who can be brought to defend it.

A detachment from every company of Grenadiers at the ſiege, with a proper number of Battalions, are ordered to join the guard of the trenches; but to prevent any diſpute about precedency or right, in making the attack, the Battalions thus ordered ſhould be thoſe who are next on command for the trenches.

A detachment of hatchet-men, with their large axes, are likewiſe ordered, that if the paſſage of the Grenadiers is obſtructed, by meeting with large paliſades, either in the covert-way, or in the intrenchments behind the breach, they may be ready to cut them down: for though the bombs and cannon from the batteries break them generally down, yet they cannot always reach them; for which reaſon there ſhould be hatchet-men ordered, for fear they ſhould be wanted.

There are likewiſe a ſufficient number of workmen ordered with tools, and others to carry the proper materials, ſuch as wool-packs, ſand-bags, gabions, faſcines, and pickets, for the making of a lodgment on the breach, if ſo ordered, or an intrenchment in the body of the outwork, to cover you from the fire of the town, and to ſecure you againſt any attempt which the beſieged ſhall make to regain it.

Engineers are commanded with the workmen, to direct them in making the proper lodgments, that no time may be loſt in the doing them.

There are always more Battalions ordered than are neceſſary for the attack, that ſome may remain as a reſerve in the trenches, which, in my opinion, ſhould be thoſe

out

out of the additional number ordered, whose tour of mounting the trenches is furthest off.

The Battalions which compose the guard of the trenches, always march after and sustain the Grenadiers, and the additional Battalions only sustain them.

The General Officers then on duty in the trenches command the attack, unless the number of troops so ordered may require a greater number of Generals than are then on duty, or one of a superior rank; in which case, the command always falls to the eldest; but unless for the reason just mentioned, the command is never taken from the Generals of the trenches.

The disposition of the troops for the attack is generally made as follows.

The Grenadiers designed for the attack, are to be posted at the head of the trenches, or that part of them which lies nearest to the work to be attacked; the particular disposition of whom is as follows:

I. A Serjeant and twelve or sixteen Grenadiers are drawn out for the forlorn hope; they are not taken from one company, but one from each of the twelve or sixteen eldest Companies; or if they consist of the troops of different nations, they are then taken in proportion to the number of Battalions of each nation.

II. A Lieutenant and thirty or forty Grenadiers formed by detachment in the same manner, to sustain the forlorn hope.

III. A Captain, two or three Lieutenants, with eighty or a hundred Grenadiers, formed also by detachment, to sustain the Lieutenant.

IV. A detachment of two hundred Grenadiers, commanded by a Major, to sustain the Captain.

V. The whole body of Grenadiers, according to seniority of Companies, or nations, under the command of Field-Officers, in proportion to their numbers. They should march as many in front as the ground they are to pass over will admit of, or the breach contain.

VI. The hatchet-men are to be posted next to the Grenadiers, and to march immediately after them.

VII.

VII. The Battalions, which compose the guard of the trenches, are posted, according to seniority, next to the hatchet-men, to sustain the Grenadiers.

VIII. The additional Battalions that are to go upon the attack, are posted next to the guard of the trenches, in order to sustain them.

IX. After the troops designed for the attack, the detachments of workmen, commanded by their Officers, are posted, that they may be ready to march, when ordered to make the lodgments, with whom the Engineers are to march to instruct them.

X. The Battalions appointed for the reserve, are posted next to the workmen; and when the others march out to the attack, they are to move up to the head of the trenches, that, if the troops which make the attack require any assistance, they may be ready to march out and sustain them, when they shall be so ordered by the General who commands the attack.

That those who make the attack may be as little exposed to the fire of the besieged as possible, all the cannon on the batteries are pointed against the several works of the town which defend the breach; on which they are to fire incessantly, during the attack, to keep the enemy from the walls.

The signal commonly given for an attack, is the throwing of a certain number of bombs into the town at the same time; but if they are thrown into the work which is to be attacked, or towards the gorge of the bastion in which the breach is made, (that being the place where the besieged intrench themselves for the defence of it) it will be of great service to those who make the attack: for as the enemy will be obliged either to quit their posts, or lie flat on the ground till the bombs are broke, it will give the Grenadiers (if they have not far to march) sufficient time to mount the breach, and attack the entrenchment without meeting with much opposition till they come there, provided the batteries fire at the same time on the defences of the town.

When

When there are more attacks than one to be made at the same time, (which, if the breaches are ready, would be exceeding proper, in order to divide the force of the garrison) each must have the same preparation and disposition made for it, unless a greater opposition is expected from the one than the other; in which case, the difference then lies in the numbers ordered for each, but not in the disposition or order of the attack.

Sham attacks are sometimes made at the same time with the real ones; but as they are intended to amuse the besieged, to oblige them to divide their troops, that those who make the real attack may meet with the less opposition, the workmen are generally omitted.

When an attack is to be made on the covert-way, the troops which are appointed for that service are generally divided into several bodies, in order to attack it at different parts at the same time. The number of workmen, with the several materials before mentioned, particularly wool-packs, are greater on these occasions; because an attack on the covert-way is generally designed to force the enemy from thence till a lodgement is made on the *Glacis*, or as it is commonly, though erroneously, called the counterscarp; for as the counterscarp is the wall of the ditch which supports the covert-way, to be lodged on the counterscarp, properly speaking, is to be lodged on the brink of the ditch; but, at present, that term is generally abused, by saying that they are on the counterscarp, when they are only at the beginning of the *Glacis*.

The most favourable time for the making of an attack, is in the day: for as the actions of every man will appear in full view, the brave, through a laudable emulation, will endeavour, at the expence of their lives, to out-do one another; and even the fearful will exert themselves, by performing their duty, rather than bear the infamous name of Coward; the fear of shame being generally more powerful than the fear of death. The batteries will be likewise of great service, by their firing

with

Chap. XVIII. *Military Discipline.*

with more certainty on the defences of the town, and the top of the breach, to keep the enemy from opposing the Grenadiers in mounting it. Besides, in the night, those who go on first will run great danger from the fire of those who sustain them; therefore an attack on an outwork, or the covert-way, is generally a little after sunset, that night may come on by the time the attack is finished, to favour them in making the necessary lodgements; but this rule will not hold good in an attack on the body of the place; for if night should come on before the town is entirely reduced to your obedience, great inconveniencies would attend both your own troops and the poor inhabitants; to avoid which, it is generally made in the forenoon.

I do not pretend, by what is mentioned in this article, to lay down certain rules; but only to give a general idea of attacks, with the usual preparation of workmen, &c. disposition of the troops, with the time of making them.

CHAP.

CHAP. XIX.

Of the method in Flanders *for the receiving and distributing of the daily orders; general detail of the army, (by which is meant the general duty to be performing by the Officers and Soldiers) with the form of a roster, or table, by which the duty of entire Battalions, and the Officers, is regulated; and a table of proportion for the detaching of private men from the whole.*

ARTICLE I.

AS the horse and foot do not interfere with one another in the detail, but have a separate one of their own, I shall therefore mention some particulars relating to the General Officers, Majors of Brigade, and the Adjutant-General, before I proceed to the orders.

Though the General Officers have not particular commissions to the horse or foot, yet their commands are distinct, as they are placed over the one or the other; for the Generals who are appointed to the horse, have the care and direction of them, and are only commanded on duty with the horse. The same rule is observed by the Generals who are appointed to the foot; so that the horse and foot do not roll together, but have each their duty a-part: however, when a detachment from each join, the eldest Officer, whether of horse or foot, commands both.

General Officers of the day are appointed for the horse and foot, each having a Lieutenant-General, and Major-General, who continue on that duty twenty-four hours, during which time they receive the orders from the General

neral in chief, and deliver them to the Generals of the horfe and foot, and Majors of Brigade of the day, as fhall be more particularly mentioned in its proper place.

The picquet is under the immediate direction of the General Officers of the day; and when it is ordered to march upon any fervice, they have the command of it: and as the picquet is not to march from the head of their feveral Regiments, but by the direction of the Lieutenant-Generals of the day, all orders relating to it fhould be immediately fent to them; for which reafon they are to be in a conftant readinefs, and not to leave the camp, but when they vifit the grand-guards and out-pofts, which lie near the army; which they generally do every morning, to know what ftate they are in, that they may acquaint the General in chief with it at orderly time, or fooner, if requifite.

All the Majors of Brigade of the foot, roll for the day to the whole body of foot, each taking it in his turn to act as fuch.

The Majors of Brigade of the horfe do the fame for the whole body of horfe.

The Majors of Brigade of the day remain on that duty twenty-four hours, during which time they keep the general detail of the whole, and regulate what each nation in particular is to furnifh to the feveral duties then ordered, and receive the orders at the head quarters from the Major-Generals of the day.

They are to fee all detachments paraded, as is explained at large in Article IV. Chap. XV. And if any difpute happens on the parade amongft the Officers about their duty, the Generals then on the fpot are to decide it according to the rules of war, or cuftom of the army; if none are prefent, it is then to be done by the Major of Brigade of the day, to which they are to fubmit; which, however, is not to be attended with any ill confequence in debarring any Officer of his right, if he can make it appear afterwards that he was wronged by the decifion.

The Majors of Brigade of each nation roll amongft themfelves for the day to their own troops, each nation

being

being to have one at the head quarters at orderly time, to receive the orders from the Major of Brigade of the day for the whole; at which time they compared and settled their books of detail with his, that they might be prepared to act for the whole in their turn; as also to see that their troops had no injustice done them in the numbers which they were to furnish. When the other Majors of Brigade received the orders from these, they compared and settled their books in the same manner, by which means they all knew the general detail; and when any of their own General Officers were to go next on duty, they sent them an account of it by their Aids-de-Camp when they came for orders, that they might be prepared for it.

The duty which was done by the Majors of Brigade of the day for the whole, in keeping of the detail, and giving of the particulars of those of each nation, was formerly performed by the Adjutant-General: which method is still continued by the *Imperialists*, their Major of Brigade of the day, or Majors of the Regiments who act as such, being only to see the guards and other detachments paraded: but the detail of the *Flanders* Army being found too great for any one person, the Majors of Brigade were ordered to execute that part day-about, from whence the Major of Brigade who kept the general detail, and distributed the orders to the others, was called Major of the day, to distinguish him from the rest: and though this took off a great deal of trouble from the Adjutant-General, yet, if he performed the other parts of his duty, he found sufficient employment.

For he is obliged to receive, and write down in his book, all orders which are given, to keep the detail both of the horse and foot, given by the Commander in Chief; and that, when the General in Chief (with whom he always remains) wants to know any thing relating to the detail of the army, he may be able to inform him.

When any orders are to be given out in the absence of the General Officers of the day, the Adjutant-General receives them from the General in Chief, and sends them

to

to the Majors of Brigade of the day for the horse and foot, if it relates to both, for their being immediately executed.

At the opening of the campaign, he is to settle with the Majors of Brigade the rosters for the several duties, as also at any other time that an alteration is required.

It is likewise his duty to inspect into the Discipline of the troops, to see that each Regiment keeps strictly to the exercise ordered.

In short, the Adjutant-General is to keep an account of every thing which passes in the Army, and attend on the General in Chief when he goes abroad, if he is not employed about some other part of his duty; but, in the day of action he is to be always near his person, to carry his orders to the Generals of the horse and foot, which is likewise the duty of his Aid-de-Camps; but when the Adjutant-General is present, and that there are any orders of consequence to be delivered to those who command the lines, he is generally sent to avoid mistakes in the giving of them; since we may reasonably suppose, that length of service, and a thorough knowledge of military affairs, were the chief motives which promoted him to that employment. The same qualifications are required in Aid-de-Camps.

Lest the out-posts should be forgot upon any sudden or unexpected march of the army, the Adjutant-General is to take care that they may be drawn in in due time, without which precaution the men on those commands may be taken or destroyed by the enemy. He is likewise to see that all the out-posts are relieved regularly, lest the Major of Brigade of the day should neglect or omit it.

ARTICLE II.

The orders are always given out at the head-quarters, and generally in the forenoon; at which time it is usual for most of the General Officers of the army to repair thither; and, as the General Officers of the day are to receive the orders from the General in Chief, they are obliged to wait upon him at that time.

The Majors of Brigade of the day for the whole, and those for each nation, are to be at the head-quarters at the same time.

The General Officers of the day, both of the horse and foot, receive the orders from the General in Chief, which should be taken in writing by the Major-Generals.

As soon as the orders are received, the Major-General of the day for the horse is to wait upon the General of the horse, and the Major-General of the day for the foot is to wait upon the General of the foot, to whom they are to deliver the orders, and to know what particular commands they have for the troops over which they are placed: but, as nothing of moment can be done but by the direction of the General in Chief, we may therefore suppose that the orders which are given by the General of the horse or foot relate only to the keeping up of discipline and order in the several corps: and what orders they receive from them, they are to acquaint the Lieutenant-Generals of the day with, and then deliver them to the Brigade-Majors of the day, to be given out with the rest.

The Adjutant-General is to deliver the orders to the Majors of Brigade of the day for the whole, and to settle with them the detail; in the doing of which they are to be very exact, that none may be ordered on duty out of their turn, or that the troops of any nation furnish more than their due proportion of Officers and private Soldiers to the several commands then ordered.

When the Majors of Brigade of the day for the whole have received the orders from, and fixed the detail with the Adjutant-General, they are to deliver them to the Majors of Brigade of the day of the several nations, with the particulars of what Officers and private men each are to furnish.

The national Majors of Brigade of the day (if I may be allowed the expression, for distinction sake) return
im-

Chap. XIX. *Military Discipline.*

immediately to their encampment, deliver the orders to the Majors of Brigade of their own troops, and settle amongst themselves what their Brigades, or Regiments, are to furnish for duties then ordered; after which, the Majors of Brigade wait upon their Major-Generals, deliver them the orders, receive their particular commands for their Brigades, and then give out the whole to the Majors of the Regiments of their respective Brigades.

The Majors wait upon their own Colonels, deliver them the orders, receive their commands for their Regiments, and then give them to the Adjutants; who wait upon their Lieutenant-Colonels, acquaint them with the orders, afterwards give them out to the orderly Serjean of each Company, and name the Officers of the Regiment who are to go on duty, with the number of private men for each Company; after which the Serjeants deliver the orders to their own Officers and Corporals (one of each Company being always present when the Serjeants receive orders, as also the Drum-Major) warn the private men, and the Drum-Major does the same by the Drummers who are to go on duty.

The Lieutenant-Generals, and Major-Generals who were not at the head-quarters at orderly-time, send their Aids-de-Camp to the Major of Brigade of the day of their own troops for the orders.

The provost-Marshal of each nation is to receive the orders in the same manner.

The Major of the Train, and the Provost-General of the army, receive orders from the Adjutant-General at the head quarters.

All orders, subsequent to those at orderly time, which the Generals of the day shall receive from the General in Chief, they are to send by their Aids-de-Camp to the Majors of Brigade of the day for the whole, that they may be immediately executed. Upon their receiving such orders, they are to send them in writing to the national Majors of Brigade of the day by their orderly Serjeants, who communicate them to the rest, and they to the Regiments of their respective Brigades by the orderly Serjeants.

When any detachment is made which is to continue out any time, it must be particularly specified in the order, that the men may be provided with ammunition-bread and pay accordingly.

Thus far I have shown how orders are received and distributed to the army, and in the following Article I shall treat of the general detail, according to the military acceptation of the word.

ARTICLE III.

All the General Officers of the foot of the same rank roll with one another, and are ordered on duty according to seniority. The same rule is observed amongst the General Officers of the horse; and, at the opening of the campaign, a list of the General Officers in the army, with the dates of their commissions, is taken by the Adjutant-General, and given by him to the Majors of Brigade of the horse and foot.

When the General Officers are to go on duty, they are always mentioned by name in publick orders.

The General of the horse, or the General of the foot, have not any fixed duty; but when a considerable Body of troops is ordered out upon any service, they are generally appointed to command them; in which case they have always one or more Lieutenant-Generals, and several Major-Generals under them, the number of whom are generally proportioned to the number of troops, or as the service on which they are to be employed may require; the particular number of men which the General Officers are to have under their command being no where fixed: for it has frequently happened that a Marshal of *France* has had under his command only ten or fifteen thousand men, and at another time an hundred thousand.

The Lieutenant-Generals, and Major-Generals, have a constant and fixed duty, as that of the day, which is explained in the foregoing Articles. Besides which, they
have

have that of commands, which is when they are ordered out with entire Battalions, or detachments from the whole; so that they have two distinct duties in the army, that of the day, and commands, which is kept by the Majors of Brigade; but when the General Officers of the day march with the picquet beyond the limits of the camp, which is the grand-guard, it passes for a command both for them and those of the picquet, and is allowed as such in the general detail.

Entire Battalions are frequently detached from the army, either for the forming of a siege, blocking up of the enemy's garrisons, securing, or covering, some part of your own country from the inroads of the enemy, or for convoys, in bringing of ammunition and provisions to the army; all of which pass for duties; but when Battalions are detached for the covering of the General's quarters, it only goes for a tour of fatigue.

As each nation had a different number of Battalions in *Flanders*, their duty was regulated by a roster; (which name, I suppose, was given it by the person who invented it) but as that of sieges was very severe service, it was made a duty a-part, that every Battalion might take their tour in process of time; so that, properly speaking, there were three distinct duties for entire Battalions. The first was sieges; the second blockades, covering of your own territories, or convoys, or commands of the like nature; and the third, covering of quarters, the detail of which was kept by the Majors of Brigade. I presume the Generals had a particular tour to sieges, as well as Battalions.

Field-Officers are not ordered on duty by name in the general orders, but by nation; each being to give as many Colonels, Lieutenant-Colonels, and Majors, as they had Battalions in the Field; so that when any were sick, wounded, or absent by leave, those of the nation, who remained, did the duty for the full complement.

They have two distinct duties in the general detail, picquet and commands, which is regulated by a roster; the form of which is exactly the same as that which made for the detaching of entire Battalions, since they are always to give an equal number with them.

In the general orders it is always said, that such a nation is to give a Colonel, such a nation a Lieutenant-Colonel, and such a Major for the picquet for such or such a wing; and when Field-Officers are ordered for commands, they are mentioned in the same manner in the general orders.

In the particular detail of each nation, the Majors of Brigade kept a list of the Field-Officers of their own troops who are present, and when it came to their turn to furnish any for the picquet or commands, they ordered them on duty by seniority, and mentioned them by name in the orders which they delivered to their own troops.

What particular duty each nation might have for their own Field-Officers, I cannot say; but that of *British*, was general Courts-Marshal, a detail of which was kept by their own Majors of Brigade.

As every nation had a different establishment of Officers to their Regiments, each nation gave therefore Captains and Subalterns only in proportion to their establishment; which duty was regulated by a roster, by taking an eighth or tenth of the Captains of each nation, and the sixteenth or twentieth of the Subalterns, and formed rosters by those numbers for the general detail of Captains and Subalterns, for whom there was only that of commands, the number which each Battalion furnished to the picquet being equal, and constantly the same; therefore the Majors of Brigade kept only that of commands.

As the Battalions of *Great Britain* were all upon the same establishment, as to their number of Officers, they all gave equally with one another, without any regard to those which had more Captains and Subalterns sick or absent than another, every Battalion being to do duty for its full complement; and I suppose the troops of every other nation observed the same rule amongst themselves.

When

Roster General for the Draughting of Battalions according to that in Flanders in 1708.

Nations	N° of Battalions belonging to each Nation	1	2	3	4	5	6	7	8	9	10	11	12	13	14	15	16	17	18	19	20	21	22	23	24	25
English	17	1-			8-		15-													39-			45-			51-
Prussians	14		2-			12-				18-											41-				49-	
Hannoverians	13			3-		11-				19-								35-					46-			
Dutch	50				4-	12-	13-	16-	17-	20-	22-	23-	25-	28-	29-	31-	33-	36-	40-	40-	42-	43-	47-	48-	50-	52-
Danes	10						14-					24-					34-					44-				

Continuation of the Roster.

Nations	N° of Battalions belonging to each Nation	26	27	28	29	30	31	32	33	34	35	36	37	38	39	40	41	42	43	44	45	46	47	48	49	50
English	17			58-			64+			72+			77+			82+			89+				97+		49-	
Prussians	14		56-			61+			68+			74+			81+			87+		91+					101-	
Hannoverians	13			59-	60-	62+		67+		71+				79+		83+			90+					99-		103-
Dutch	50	54-	57-			63+	65+		69+	73+	73+	75+	78+	80+		84+	85+	88+		92+	93+	96+	98-	100-		104-
Danes	10	55-					66+					76+					86+									

Chap. XIX.   *Military Difcipline.*   343

When any *Britifh* Captains and Subalterns were appointed for commands by the general orders, their own Majors of Brigade regulated amongft themfelves what Battalions were to furnifh them; and, in the giving of the orders to their own troops they mentioned the Regiments by name who were to furnifh Captains, and who Subalterns.

The particular duty for the Captains of the *Englifh* Battalions was that of Courts-Marfhal, which detail was kept by their own Majors of Brigade.

Every Battalion in the army, whether ftrong or weak, gives an equal number of private Men to all detachments; for the ufe of which, I have hereunto annexed a table of proportion, that no time may be loft, or a miftake made by a wrong calculation.

Thefe were the eftablifhed rules for the detail in *Flanders*, both as it regarded the whole, and that of each nation in particular; and in the next place I fhall endeavour to fhew, in as clear a manner as I can, the form and ufe of a rofter, or table, for regulating the duty of an army, which is compofed of the troops of different Princes, whofe number of Regiments, or Officers, are unequal.

The rofter is ufed by the horfe as well as the foot, for the regulating of the duty of entire Squadrons, which are more frequently detached than Battalions.

Explanation of the following rofter.

In the firft column are the names of the feveral nations; and in the fecond, the number of Battalions which each had; and, as the higheft number was fifty, which belonged to the *Dutch*, fifty columns more are added, which makes fifty fquares oppofite to each nation; but as the *Englifh* have but feventeen Battalions, and being only to give in proportion to that number, all the fquares but feventeen are filled up: the fame is obferved by thofe of *Pruffia*, *Hanover*, and *Denmark*, each having no more blank fquares left than they have Battalions.

The method of placing the blank squares at a distance, and filling up of those between them, may be seen by the plan; but the reason for dividing of them in this manner will appear very plain, when the method of detaching of Battalions, by the roster, is known.

As 17 to 50 is almost 1 to 3, the dividing of the blank squares opposite to the *English*, is very regular and easy; as 10 to 50 is 1 to 5, which is the *Danes*, theirs is quite regular; but those of *Prussia* and *Hanover*, not bearing so near a proportion, the number of the filled-up squares between the blanks, will, of course, vary.

All the Columns are numbered on the top from 1 to 50; and, as the blank squares in the several columns are supposed to be Battalions, I have numbered them from 1 to 104, as they are to be detached the one after the other; which shews the method of detaching of them in so clear a manner, that it must be conceived at first view, and will enable any one to form rosters for any number, and save me the trouble of adding more plans of this nature.

But in order to see how the proportion answers, let us suppose three sieges to be undertaken at different times, to each of which thirty Battalions are to be detached.

The first thirty Battalions begins with column 1, and ends with column 14, the blanks in which columns I have marked with a point, or stop, to distinguish them from the rest.

The second begins with column 15, and ends with column 29, the blanks in which are marked with a stroke, thus (—). The third siege begins with column 30, and ends with column 43, the blanks in those columns are marked with a cross, thus (†).

The following table will show the number of Battalions each nation is to furnish to the several sieges, and what number remains undetached of the whole; and as calculations of this nature will not admit of fractions, it is impossible to bring the proportions nearer, or invent a more proper method, for the purpose, than this.

Chap. XIX. *Military Discipline.*

Nations.	Number of Battalions at the first siege.	Number of Battalions at the second siege.	Number of Battalions at the third siege.	Number of Battalions remained undetached.	Number of Battalions which belonged to each nation.
English	5	5	5	2	17
Prussians	4	4	4	2	14
Hanoverians	4	3	4	2	13
Dutch	14	15	14	7	50
Danes	3	3	3	1	10
Total	30	30	30	14	104

Explanation of the following table.

As the table is carried no further than from ten Battalions to one hundred and nine, and from two men a Battalion to seventy-one, there may be an objection for its not being more complete; since detachments of a greater number of men than seventy-one a Regiment are often commanded, particularly at sieges, and in a garrison: but, as my principal design is only to shew the use of the table, if that end is complied with, I think I have fully answered the purpose; for when the method is known, every one may make a table of proportion to as high a number as they please for their own use.

Every leaf, or two sides, completes the table of ten Battalions as far as the calculation is carried.

The first leaf begins with ten Battalions, and ends with nineteen: the second leaf begins with twenty Battalions, and ends with twenty-nine, and so with the rest, to one hundred and nine Battalions, as may be seen by the figures on the top.

The first side of every leaf begins with two men a Battalion, and ends with thirty-six; the second side of
every

every one begins with thirty-seven men a Battalion, and ends with seventy-one, as may be seen by the figures in the margin, or first column.

The use of the table is as follows.

When a number of men are to be detached, and that you want to know the proportion which each Regiment is to give, you must find in the top the number which the Battalions in the army consists of. After that carry your eye down the column till you find the number ordered, and then trace the line, in which the number stands, till you come to the margin, or first column, and the figures there shew you the number which each Battalion is to furnish; but as it will not always happen that you can find in the column the exact number ordered, but that some will be wanting, or exceed it, you must stop at that which comes the nearest to it, but always less than the number required. The men thus wanting are called odd men, because they do not come exactly to a man a Battalion. When this is the case, you then order as many Battalions as there are men wanting to give each a man more than what are mentioned in the margin, an account of which is kept by the Majors of Brigade, that every Battalion may furnish odd men in their turn.

But lest it should not be fully comprehended by the above explanation, a few examples, I believe, will make it indisputably so.

Suppose a detachment is to be made of five hundred and fifty men from sixteen Battalions; you must find out the column on the top number 16; then look down the column, till you come to five hundred and forty-four, (which is the nearest you can come to the number ordered without exceeding it, which you are never to do) and you will find the figures in the margin opposite to that number to be thirty-four, which is thirty-four men a Battalion; but as thirty four men a Battalion, makes only five hundred and forty-four, and that five

hundred

hundred and fifty are required, six Battalions must therefore give thirty-five men each, and the other ten Battalions only thirty-four men each.

Let us suppose further, that sixteen hundred men are ordered from twenty-five Battalions. You must find out the number twenty-five on the top, and look down that column; but, as the highest number on the first side of that leaf is nine hundred, you must turn over and look down column twenty-five on the back of it till you come to sixteen hundred, and you will find the figures opposite to it in the margin to be sixty-four, which is the number each Battalion is to give.

I am persuaded, that I need not trouble the reader with a further explanation, since it must be thoroughly understood by what is already said of it.

As detachments from the horse are made from the Squadrons, as the foot are from the Battalions, the table will be as useful to the cavalry as it is to the infantry, by putting in the word Squadron instead of Battalion.

Table

Table of Proportion for the detaching of private Men.

Numb. of Men to be detached from each Battalion.	Number of Battalions from which the Men are to be detached.									
	10	11	12	13	14	15	16	17	18	19
2	20	22	24	26	28	30	32	34	36	38
3	30	33	36	39	42	45	48	51	54	57
4	40	44	48	52	56	60	64	68	72	76
5	50	55	60	65	70	75	80	85	90	95
6	60	66	72	78	84	90	96	102	108	114
7	70	77	84	91	98	105	112	119	126	133
8	80	88	96	104	112	120	128	136	144	152
9	90	99	108	117	126	135	144	153	162	171
10	100	110	120	130	140	150	160	170	180	190
11	110	121	132	143	154	165	176	187	198	209
12	120	132	144	156	168	180	192	204	216	228
13	130	143	156	169	182	195	208	221	234	247
14	140	154	168	182	196	210	224	238	252	266
15	150	165	180	195	210	225	240	255	270	285
16	160	176	192	208	224	240	256	272	288	304
17	170	187	204	221	238	255	272	289	306	323
18	180	198	266	234	252	270	288	306	324	342
19	190	209	228	247	266	285	304	323	342	361
20	200	220	240	260	280	300	320	340	360	380
21	210	231	252	273	294	315	336	357	378	399
22	220	242	264	286	308	330	352	374	396	418
23	230	253	276	299	322	345	368	391	414	437
24	240	264	288	312	336	360	384	408	432	456
25	250	275	300	325	350	375	400	425	450	475
26	260	286	312	338	364	390	416	442	468	494
27	270	297	324	351	378	405	432	459	486	513
28	280	308	336	364	392	420	448	476	504	532
29	290	319	348	377	406	435	464	493	522	551
30	300	330	360	390	420	450	480	510	540	570
31	310	341	372	403	434	465	496	527	558	589
32	320	352	384	416	448	480	512	544	576	608
33	330	363	396	429	462	495	528	561	594	627
34	340	374	408	442	476	510	544	578	612	646
35	350	385	420	455	490	525	560	595	630	665
36	360	396	432	468	504	540	576	612	648	684

Battalions from which the number

Number of Men to be detached	10	11	12	13	14	15	16	17	18	19
37	370	407	444	481	518	555	592	629	666	703
38	380	418	456	494	532	570	608	646	684	722
39	390	429	468	507	546	585	624	663	702	741
40	400	440	480	520	560	600	640	680	720	760
41	410	451	492	533	574	615	656	697	738	779
42	420	462	504	546	588	630	672	714	756	798
43	430	473	516	559	602	645	688	731	774	817
44	440	484	528	572	616	660	704	748	792	836
45	450	495	540	585	630	675	720	765	810	855
46	460	506	552	598	644	690	736	782	828	874
47	470	517	564	611	658	705	752	799	846	893
48	480	528	576	624	672	720	768	816	864	912
49	490	539	588	637	686	735	784	833	882	931
50	500	550	600	650	700	750	800	850	900	950
51	510	561	612	663	714	765	816	867	918	969
52	520	572	624	676	728	780	832	884	936	988
53	530	583	636	689	742	795	848	901	954	1007
54	540	594	648	702	756	810	864	918	972	1026
55	550	605	660	715	770	825	880	935	990	1045
56	560	616	672	728	784	840	896	952	1008	1064
57	570	627	684	741	798	855	912	969	1026	1083
58	580	638	696	754	812	870	928	986	1044	1102
59	590	649	708	767	826	885	944	1003	1062	1121
60	600	660	720	780	840	900	960	1020	1080	1140
61	610	671	732	793	854	915	976	1037	1098	1159
62	620	682	744	806	868	930	992	1054	1116	1178
63	630	693	756	819	882	945	1008	1071	1134	1197
64	640	704	768	832	896	960	1024	1088	1152	1216
65	650	715	780	845	910	975	1040	1105	1170	1235
66	660	726	792	858	924	990	1056	1122	1188	1254
67	670	737	804	871	938	1005	1072	1139	1206	1273
68	680	748	816	884	952	1020	1088	1156	1224	1292
69	690	759	828	897	966	1035	1104	1173	1242	1311
70	700	770	840	910	980	1050	1120	1190	1260	1330
71	710	781	852	923	994	1065	1136	1207	1278	1349

in the margin are to be detached.

Number of Men to be detached.	20	21	22	23	24	25	26	27	28	29
2	40	42	44	46	48	50	52	54	56	58
3	60	63	66	69	72	75	78	81	84	87
4	80	84	88	92	96	100	104	108	112	116
5	100	105	110	115	120	125	130	134	140	145
6	120	126	132	138	144	150	156	162	168	174
7	140	147	154	161	168	175	182	189	196	203
8	160	168	176	184	192	200	208	216	224	232
9	180	189	198	207	216	225	234	243	252	261
10	200	210	220	230	240	250	260	270	280	290
11	220	231	242	253	264	275	286	297	308	319
12	240	252	264	276	288	300	312	324	336	348
13	260	273	286	299	312	325	338	351	364	377
14	280	294	308	322	336	350	364	378	392	406
15	300	315	330	345	360	375	390	405	420	435
16	320	336	352	368	384	400	416	432	448	464
17	340	357	374	391	408	425	442	459	476	493
18	360	378	396	414	432	450	468	486	504	522
19	380	399	418	437	456	475	494	513	532	551
20	400	420	440	460	480	500	520	540	560	580
21	420	441	462	483	504	525	546	567	588	609
22	440	462	484	506	528	550	572	594	616	638
23	460	483	506	529	552	575	598	621	644	667
24	480	504	528	552	576	600	624	648	672	696
25	500	525	550	575	600	625	650	675	700	725
26	520	546	572	598	624	650	676	702	728	754
27	540	567	594	621	648	675	702	729	756	783
28	560	588	616	644	672	700	728	756	784	812
29	580	609	638	667	696	725	754	783	812	841
30	600	630	660	690	720	750	780	810	840	870
31	620	651	682	713	744	775	806	837	868	899
32	640	672	704	736	768	800	832	864	896	928
33	660	693	726	759	792	825	858	891	924	957
34	680	714	748	782	816	850	884	918	952	986
35	700	735	770	805	840	875	910	945	980	1015
36	720	756	792	828	864	900	936	972	1008	1044

Battalions from which the number

Number of Men to be detached.	20	21	22	23	24	25	26	27	28	29
37	740	777	814	851	888	925	962	999	1036	1073
38	760	798	836	874	912	950	988	1026	1064	1102
39	780	819	858	897	936	975	1014	1053	1092	1131
40	800	840	880	920	960	1000	1040	1080	1120	1160
41	820	861	902	943	984	1025	1066	1107	1148	1189
42	840	882	924	966	1008	1050	1092	1134	1176	1218
43	860	903	946	989	1032	1075	1118	1161	1204	1247
44	880	924	968	1012	1056	1100	1144	1188	1232	1276
45	900	945	990	1035	1080	1125	1170	1215	1260	1305
46	920	966	1012	1058	1104	1150	1196	1242	1288	1334
47	940	987	1034	1081	1128	1175	1222	1269	1316	1363
48	960	1008	1056	1104	1152	1200	1248	1296	1344	1392
49	980	1029	1078	1127	1176	1225	1274	1323	1372	1421
50	1000	1050	1100	1150	1200	1250	1300	1350	1400	1450
51	1020	1071	1122	1173	1224	1275	1326	1377	1428	1479
52	1040	1092	1144	1196	1248	1300	1352	1404	1456	1508
53	1060	1113	1166	1219	1272	1325	1378	1431	1484	1537
54	1080	1134	1188	1242	1296	1350	1404	1458	1512	1566
55	1100	1155	1210	1265	1320	1375	1430	1485	1540	1595
56	1120	1176	1232	1288	1344	1400	1456	1512	1568	1624
57	1140	1197	1254	1311	1368	1425	1482	1539	1596	1653
58	1160	1218	1276	1334	1392	1450	1508	1566	1624	1682
59	1180	1239	1298	1357	1416	1475	1534	1593	1652	1711
60	1200	1260	1320	1380	1440	1500	1560	1620	1680	1740
61	1220	1281	1342	1403	1464	1525	1586	1647	1708	1769
62	1240	1302	1364	1426	1488	1550	1612	1674	1736	1798
63	1260	1323	1386	1449	1512	1575	1638	1701	1764	1827
64	1280	1344	1408	1472	1536	1600	1664	1728	1792	1856
65	1300	1365	1430	1495	1560	1625	1690	1755	1820	1885
66	1320	1386	1452	1518	1584	1650	1716	1782	1848	1914
67	1340	1407	1474	1541	1608	1675	1742	1809	1876	1943
68	1360	1428	1496	1564	1632	1700	1768	1836	1904	1972
69	1380	1449	1518	1587	1656	1725	1794	1863	1932	2001
70	1400	1470	1540	1610	1680	1750	1820	1890	1960	2030
71	1420	1491	1562	1633	1704	1775	1846	1917	1988	2059

in the margin are to be detached.

Number of Men to be detached.	30	31	32	33	34	35	36	37	38	39
2	60	62	64	66	68	70	72	74	76	78
3	90	93	96	99	102	105	108	111	114	117
4	120	124	128	132	136	140	144	148	152	156
5	150	155	160	165	170	175	180	185	190	195
6	180	186	192	198	204	210	216	222	228	234
7	210	217	224	231	238	245	252	259	266	273
8	240	248	256	264	272	280	288	296	304	312
9	270	279	288	297	306	315	324	333	342	351
10	300	310	320	330	340	350	360	370	380	390
11	330	341	352	363	374	385	396	407	418	429
12	360	372	384	396	408	420	432	444	456	468
13	390	403	416	429	442	455	468	481	494	507
14	420	434	448	462	476	490	504	518	532	546
15	450	465	480	495	510	525	540	555	570	585
16	480	496	512	528	544	560	576	592	608	624
17	510	527	544	561	578	595	612	629	646	663
18	540	558	576	594	612	630	648	666	684	702
19	570	589	608	627	646	665	684	703	722	741
20	600	620	640	660	680	700	720	740	760	780
21	630	651	672	693	714	735	756	777	798	819
22	660	682	704	726	748	770	792	814	836	858
23	690	713	736	759	782	805	828	851	874	897
24	720	744	768	792	816	840	864	888	912	936
25	750	775	800	825	850	875	900	925	950	975
26	780	806	832	858	884	910	936	962	988	1014
27	810	837	864	891	918	945	972	999	1026	1053
28	840	868	896	924	952	980	1008	1036	1064	1092
29	870	899	928	957	986	1015	1044	1073	1102	1131
30	900	930	960	990	1020	1050	1080	1110	1140	1170
31	930	961	992	1023	1054	1085	1116	1147	1178	1209
32	960	992	1024	1056	1088	1120	1152	1184	1216	1248
33	990	1023	1056	1089	1122	1155	1188	1221	1254	1287
34	1020	1054	1088	1122	1156	1190	1224	1258	1292	1326
35	1050	1085	1120	1155	1190	1225	1260	1295	1330	1365
36	1080	1116	1152	1188	1224	1260	1296	1332	1368	1404

Battalions from which the number

Number to Men of be detached.	30	31	32	33	34	35	36	37	38	39
37	1110	1147	1184	1221	1258	1295	1332	1369	1406	1443
38	1140	1178	1216	1254	1292	1330	1368	1406	1444	1482
39	1170	1209	1248	1287	1326	1365	1404	1443	1482	1521
40	1200	1240	1280	1320	1360	1400	1440	1480	1520	1560
41	1230	1271	1312	1353	1394	1435	1476	1517	1558	1599
42	1260	1302	1344	1386	1428	1470	1512	1554	1596	1638
43	1290	1333	1376	1419	1462	1505	1548	1591	1634	1677
44	1320	1364	1408	1452	1496	1540	1584	1628	1672	1716
45	1350	1395	1440	1480	1530	1575	1620	1665	1710	1755
46	1380	1426	1472	1518	1564	1610	1656	1702	1748	1794
47	1410	1457	1504	1551	1598	1645	1692	1739	1786	1833
48	1440	1488	1536	1584	1632	1680	1728	1776	1824	1872
49	1470	1519	1568	1617	1676	1715	1764	1813	1862	1911
50	1500	1550	1600	1650	1700	1750	1800	1850	1900	1950
51	1530	1581	1632	1683	1734	1785	1836	1887	1938	1989
52	1560	1612	1664	1716	1768	1820	1872	1924	1976	2028
53	1590	1643	1697	1749	1802	1855	1908	1961	2014	2067
54	1620	1674	1728	1782	1836	1890	1944	1998	2052	2106
55	1650	1705	1760	1815	1870	1925	1980	2035	2090	2145
56	1680	1736	1792	1848	1904	1960	2016	2072	2128	2184
57	1710	1767	1824	1881	1938	1995	2052	2109	2166	2223
58	1740	1798	1856	1914	1972	2030	2088	2146	2204	2262
59	1770	1829	1888	1947	2006	2065	2124	2183	2242	2301
60	1800	1860	1920	1980	2040	2100	2160	2220	2280	2340
61	1830	1891	1952	2013	2074	2135	2196	2257	2318	2379
62	1860	1922	1984	2046	2108	2170	2232	2294	2356	2418
63	1890	1953	2016	2079	2142	2205	2268	2331	2394	2457
64	1920	1984	2048	2112	2176	2240	2304	2368	2432	2496
65	1950	2015	2080	2145	2210	2275	2340	2405	2470	2535
66	1980	2046	2112	2178	2244	2310	2376	2442	2508	2574
67	2010	2077	2144	2211	2278	2345	2412	2479	2546	2613
68	2040	2108	2176	2244	2312	2380	2448	2516	2584	2652
69	2070	2139	2208	2277	2346	2415	2484	2553	2622	2691
70	2100	2170	2240	2310	2380	2450	2520	2590	2660	2730
71	2130	2201	2272	2343	2414	2485	2556	2627	2698	2769

in the margin are to be detached.

Number of Men to be detached.	40	41	42	43	44	45	46	47	48	49
2	80	82	84	86	88	90	92	94	96	98
3	120	123	126	129	132	135	138	141	144	147
4	160	164	168	172	176	180	184	188	192	196
5	200	205	210	215	220	225	230	235	240	245
6	240	246	252	258	264	270	276	282	288	294
7	280	287	294	301	308	315	322	329	336	343
8	320	328	336	344	352	360	368	376	384	392
9	360	369	378	387	396	405	414	423	432	441
10	400	410	420	430	440	450	460	470	480	490
11	440	451	462	473	484	495	506	517	528	539
12	480	492	504	516	528	540	552	564	576	588
13	520	533	546	559	572	585	598	611	624	637
14	560	574	588	602	616	630	644	658	672	686
15	600	615	630	645	660	675	690	705	720	735
16	640	656	672	688	704	720	736	752	768	784
17	680	697	714	731	748	765	782	799	816	833
18	720	738	756	774	792	810	828	846	864	882
19	760	779	798	817	836	855	874	893	912	931
20	800	820	840	860	880	900	920	940	960	980
21	840	861	882	903	924	945	966	987	1008	1029
22	880	902	924	946	968	990	1012	1034	1056	1078
23	920	943	966	989	1012	1035	1058	1081	1104	1127
24	960	984	1008	1032	1056	1080	1104	1128	1152	1176
25	1000	1025	1050	1075	1100	1125	1150	1175	1200	1225
26	1040	1066	1092	1118	1144	1170	1196	1222	1248	1274
27	1080	1107	1134	1161	1188	1215	1242	1269	1296	1323
28	1120	1148	1176	1204	1232	1260	1288	1316	1344	1372
29	1160	1189	1218	1247	1276	1305	1334	1363	1392	1421
30	1200	1230	1260	1290	1320	1350	1380	1410	1440	1470
31	1240	1271	1302	1333	1364	1395	1426	1457	1488	1519
32	1280	1312	1344	1376	1408	1440	1472	1504	1536	1568
33	1320	1353	1386	1419	1452	1485	1518	1551	1584	1617
34	1360	1394	1428	1462	1496	1530	1564	1598	1632	1666
35	1400	1435	1470	1505	1540	1575	1610	1645	1680	1715
36	1440	1476	1512	1548	1584	1620	1656	1692	1728	1764

Battalions from which the number

Number of Men to be detached.	40	41	42	43	44	45	46	47	48	49
37	1480	1517	1554	1591	1628	1665	1702	1739	1776	1813
38	1520	1558	1596	1634	1672	1710	1748	1786	1824	1862
39	1560	1599	1638	1677	1716	1755	1794	1833	1872	1911
40	1600	1640	1680	1720	1760	1800	1840	1880	1920	1960
41	1640	1681	1722	1763	1804	1845	1886	1927	1968	2009
42	1680	1722	1764	1806	1848	1890	1932	1974	2016	2058
43	1720	1763	1806	1849	1892	1935	1978	2021	2064	2107
44	1760	1804	1848	1892	1936	1980	2024	2068	2112	2156
45	1800	1845	1890	1935	1980	2025	2070	2115	2160	2205
46	1840	1886	1932	1978	2024	2070	2116	2162	2208	2254
47	1880	1927	1974	2021	2068	2115	2162	2209	2256	2303
48	1920	1968	2016	2064	2112	2160	2208	2256	2304	2352
49	1960	2009	2058	2107	2156	2205	2254	2303	2352	2401
50	2000	2050	2100	2150	2200	2250	2300	2350	2400	2450
51	2040	2091	2142	2193	2244	2295	2346	2397	2448	2499
52	2080	2132	2184	2236	2288	2340	2392	2444	2496	2548
53	2120	2173	2226	2279	2332	2385	2438	2491	2544	2597
54	2160	2214	2268	2322	2376	2430	2484	2538	2592	2646
55	2200	2255	2310	2365	2420	2475	2530	2585	2640	2695
56	2240	2296	2352	2408	2464	2520	2576	2632	2688	2744
57	2280	2337	2394	2451	2508	2565	2622	2679	2736	2793
58	2320	2378	2436	2494	2552	2610	2668	2726	2784	2842
59	2360	2419	2478	2537	2596	2655	2714	2773	2832	2891
60	2400	2460	2520	2580	2640	2700	2760	2820	2880	2940
61	2440	2501	2562	2623	2684	2745	2806	2867	2928	2989
62	2480	2542	2604	2666	2728	2790	2852	2914	2976	3038
63	2520	2583	2646	2709	2772	2835	2898	2961	3024	3087
64	2560	2624	2688	2752	2816	2880	2944	3008	3072	3136
65	2600	2665	2730	2795	2860	2925	2990	3055	3120	3185
66	2640	2706	2772	2838	2904	2970	3036	3102	3168	3234
67	2680	2747	2814	2881	2948	3015	3082	3149	3216	3283
68	2720	2788	2856	2924	2992	3060	3128	3196	3264	3332
69	2760	2829	2898	2967	3036	3105	3174	3243	3312	3381
70	2800	2870	2940	3010	3080	3150	3220	3290	3360	3430
71	2840	2911	2982	3053	3124	3195	3266	3337	3408	3479

in the margin are to be detached.

Number of Men to be detached.	50	51	52	53	54	55	56	57	58	59
2	100	102	104	106	108	110	112	114	116	118
3	150	153	156	159	162	165	168	171	174	177
4	200	204	208	212	216	220	224	228	232	236
5	250	255	260	265	270	275	289	285	290	295
6	300	306	312	318	324	330	336	342	348	354
7	350	357	364	371	378	385	392	399	406	413
8	400	408	416	424	432	440	448	456	464	472
9	450	459	468	477	486	495	504	513	522	531
10	500	510	520	530	540	550	560	570	580	590
11	550	561	572	583	594	605	616	627	638	649
12	600	612	624	636	648	660	672	684	696	708
13	650	663	676	689	702	715	728	741	754	767
14	700	714	728	742	756	770	784	798	812	826
15	750	765	780	795	810	825	840	855	870	885
16	800	816	832	848	864	880	896	912	928	944
17	850	867	884	901	918	935	952	969	986	1003
18	900	918	936	954	972	990	1008	1026	1044	1062
19	950	969	988	1007	1026	1045	1064	1083	1102	1121
20	1000	1020	1040	1060	1080	1100	1120	1140	1160	1180
21	1050	1071	1092	1113	1134	1115	1176	1197	1218	1239
22	1100	1122	1144	1166	1188	1210	1232	1254	1276	1298
23	1150	1173	1196	1219	1242	1265	1288	1311	1334	1357
24	1200	1224	1248	1272	1296	1320	1344	1368	1392	1416
25	1250	1275	1300	1325	1350	1375	1400	1425	1450	1475
26	1300	1326	1352	1378	1404	1430	1456	1482	1508	1534
27	1350	1377	1404	1431	1458	1485	1512	1539	1566	1593
28	1400	1428	1456	1484	1512	1540	1568	1569	1624	1652
29	1450	1479	1508	1537	1566	1595	1624	1653	1682	1711
30	1500	1530	1560	1590	1620	1650	1680	1710	1740	1770
31	1550	1581	1612	1643	1674	1705	1736	1767	1798	1829
32	1600	1632	1664	1696	1728	1760	1792	1824	1856	1888
33	1650	1683	1716	1749	1782	1815	1848	1881	1914	1947
34	1700	1734	1768	1802	1836	1870	1904	1938	1972	2006
35	1750	1785	1820	1855	1890	1925	1960	1995	2030	2065
36	1800	1836	1872	1908	1944	1908	2016	2052	2088	2124

Battalions from which the number

Number of Men to be detached.	50	51	52	53	54	55	56	57	58	59
37	1850	1887	1924	1961	1998	2035	2072	2109	2146	2183
38	1900	1938	1976	2014	2052	2090	2128	2166	2204	2242
39	1950	1989	2028	2067	2106	2145	2184	2223	2262	2301
40	2000	2040	2080	2120	2161	2200	2240	2280	2320	2360
41	2050	2091	2132	2173	2214	2255	2296	2337	2378	2419
42	2100	2142	2184	2226	2268	2310	2352	2394	2436	2478
43	2150	2193	2236	2279	2322	2365	2408	2451	2494	2537
44	2200	2244	2288	2332	2376	2420	2464	2508	2552	2596
45	2250	2295	2340	2385	2430	2475	2520	2565	2610	2655
46	2300	2346	2392	2438	2484	2530	2576	2622	2668	2714
47	2350	2397	2444	2491	2538	2585	2632	2679	2726	2773
48	2400	2448	2496	2544	2592	2640	2688	2736	2784	2832
49	2450	2499	2548	2597	2646	2695	2744	2793	2842	2891
50	2500	2550	2600	2650	2700	2750	2800	2850	2900	2950
51	2550	2601	2652	2703	2754	2805	2856	2907	2958	3009
52	2600	2652	2704	2756	2808	2860	2912	2964	3016	3068
53	2650	2703	2756	2809	2862	2915	2968	3021	3074	3127
54	2700	2754	2808	2862	2916	2970	3024	3078	3132	3186
55	2750	2805	2860	2915	2970	3025	3080	3135	3190	3245
56	2800	2856	2912	2968	3024	3080	3136	3192	3248	3304
57	2850	2907	2964	3021	3078	3135	3192	3249	3306	3363
58	2900	2958	3016	3074	3132	3190	3248	3306	3364	3422
59	2950	3009	3068	3127	3186	3245	3304	3363	3422	3481
60	3000	3060	3120	3180	3240	3300	3360	3420	3480	3540
61	3050	3111	3172	3233	3294	3355	3416	3477	3538	3599
62	3100	3162	3224	3286	3348	3410	3472	3534	3596	3658
63	3150	3213	3276	3339	3402	3465	3528	3591	7654	3717
64	3200	3264	3328	3392	3456	3520	3584	3648	3712	3776
65	3250	3315	3380	3445	3510	3575	3640	3705	3770	3835
66	3300	3366	3432	3498	3564	3630	3696	3762	3828	3894
67	3350	3417	3484	3551	3618	3685	3782	3819	3886	3953
68	3400	3468	3536	3604	3672	3740	3808	3876	3944	4012
69	3450	3519	3588	3657	3726	3795	3884	3933	4002	4071
70	3500	3570	3640	3710	3780	3850	3920	3990	4060	4130
71	3550	3621	3692	3766	3834	3905	3976	4047	4118	4189

in the margin are to be detached.

Number of Men to be detached	60	61	62	63	64	65	66	67	68	69
2	120	122	124	126	128	130	132	134	136	138
3	180	183	186	189	192	195	198	201	204	207
4	240	244	248	252	256	260	264	268	272	276
5	300	305	310	315	320	325	330	335	340	345
6	360	366	372	378	384	390	396	402	408	414
7	420	427	434	441	448	455	462	469	476	483
8	480	488	496	504	512	520	528	536	544	552
9	540	549	558	567	576	585	594	603	612	621
10	600	610	620	630	640	650	660	670	680	690
11	660	671	682	693	704	715	726	737	748	759
12	720	732	744	756	768	780	792	804	816	828
13	780	793	806	819	832	845	858	871	884	897
14	840	854	868	882	896	910	924	938	952	966
15	900	915	930	945	960	975	990	1005	1020	1035
16	960	976	992	1008	1024	1040	1056	1072	1088	1104
17	1020	1037	1054	1071	1088	1105	1122	1139	1156	1173
18	1080	1098	1116	1134	1152	1170	1188	1206	1224	1242
19	1140	1159	1178	1197	1216	1235	1254	1273	1292	1311
20	1200	1220	1240	1260	1280	1300	1320	1340	1360	1380
21	1260	1281	1302	1323	1344	1365	1386	1407	1428	1449
22	1320	1342	1364	1386	1408	1430	1452	1474	1496	1518
23	1380	1403	1426	1449	1472	1495	1518	1541	1564	1587
24	1440	1464	1488	1512	1536	1560	1584	1608	1632	1656
25	1500	1525	1550	1575	1600	1625	1650	1675	1700	1725
26	1560	1586	1612	1638	1664	1690	1716	1742	1768	1794
27	1620	1647	1674	1701	1728	1755	1782	1809	1836	1863
28	1680	1708	1736	1764	1792	1820	1848	1876	1904	1932
29	1740	1769	1798	1827	1856	1885	1914	1943	1972	2001
30	1800	1830	1860	1890	1920	1950	1980	2010	2040	2070
31	1860	1891	1922	1953	1984	2015	2046	2077	2108	2139
32	1920	1952	1984	2016	2048	2080	2112	2144	2176	2208
33	1980	2013	2046	2079	2112	2145	2178	2211	2244	2277
34	2040	2074	2108	2142	2176	2210	2244	2278	2312	2346
35	2100	2135	2170	2205	2240	2275	2310	2345	2380	2415
36	2160	2196	2232	2268	2304	2340	2376	2412	2448	2484

Battalions from which the number

Number of Men to be detached.	60	61	62	63	64	65	66	67	68	69
37	2220	2257	2294	2331	2368	2405	2442	2479	2516	2553
38	2280	2318	2356	2394	2432	2470	2508	2546	2584	2622
39	2340	2379	2418	2457	2496	2535	2574	2613	2652	2691
40	2400	2440	2480	2520	2560	2600	2640	2680	2720	2760
41	2460	2501	2542	2583	2624	2665	2706	2747	2788	2829
42	2520	2562	2604	2646	2688	2730	2772	2814	2856	2898
43	2580	2623	2666	2709	2752	2795	2838	2881	2924	2967
44	2640	2684	2728	2772	2816	2860	2905	2948	2992	3036
45	2700	2745	2790	2835	2880	2925	2970	3015	3060	3105
46	2760	2806	2852	2898	2944	2990	3036	3082	3128	3174
47	2820	2867	2914	2961	3008	3055	3102	3149	3196	3243
48	2880	2928	2976	3024	3072	3120	3168	3216	3264	3312
49	2940	2989	3038	3087	3136	3185	3234	3283	3332	3381
50	3000	3050	3100	3150	3200	3250	3300	3350	3400	3450
51	3060	3111	3162	3213	3264	3315	3366	3417	3468	3519
52	3120	3172	3224	3276	3328	3380	3432	3484	3536	3588
53	3180	3233	3286	3339	3392	3445	3498	3551	3604	3657
54	3240	3294	3348	3402	3456	3510	3564	3618	3672	3726
55	3300	3355	3410	3465	3520	3575	3630	3685	3740	3795
56	3360	3416	3472	3528	3584	3640	3696	3752	3808	3864
57	3420	3477	3534	3591	3648	3705	3762	3819	3876	3933
58	3480	3538	3596	3654	3712	3770	3828	3886	3944	4002
59	3540	3599	3658	3717	3776	3835	3894	3953	4012	4071
60	3600	3660	3720	3780	3840	3900	3960	4020	4080	4140
61	3660	3721	3782	3843	3904	3965	4026	4087	4148	4209
62	3720	3782	3844	3906	3968	4030	4092	4154	4216	4278
63	3780	3843	3906	3969	4032	4095	4158	4221	4284	4347
64	3840	3904	3968	4032	4096	4160	4224	4288	4352	4416
65	3900	3965	4030	4095	4160	4225	4290	4355	4420	4485
66	3960	4026	4092	4158	4224	4290	4356	4422	4488	4554
67	4020	4087	4154	4221	4288	4355	4422	4489	4556	4623
68	4080	4148	4216	4284	4352	4420	4488	4556	4624	4692
69	4140	4209	4278	4347	4416	4485	4554	4623	4692	4761
70	4200	4270	4340	4410	4480	4550	4620	4690	4760	4830
71	4260	4331	4402	4473	4544	4615	4686	4757	4828	4899

in the margin are to be detached.

Number of Men to be detached.	70	71	72	73	74	75	76	77	78	79
2	140	142	144	146	148	150	152	154	156	158
3	210	213	216	219	222	225	228	231	234	237
4	280	284	288	292	296	300	304	308	312	316
5	350	355	360	365	370	375	380	385	390	395
6	420	426	432	438	444	450	456	462	468	474
7	490	497	504	511	518	525	532	539	546	553
8	560	568	576	584	592	600	608	616	624	632
9	630	639	648	657	666	675	684	693	702	711
10	700	710	720	730	740	750	760	770	780	790
11	770	781	792	803	814	825	836	847	858	869
12	840	852	864	876	888	900	912	924	936	948
13	910	923	936	949	962	975	988	1001	1014	1027
14	980	994	1008	1022	1036	1050	1064	1078	1092	1106
15	1050	1065	1080	1095	1110	1125	1140	1155	1170	1185
16	1120	1136	1152	1168	1184	1200	1216	1232	1248	1264
17	1190	1207	1224	1241	1258	1275	1292	1309	1326	1343
18	1260	1278	1296	1314	1332	1350	1368	1386	1404	1422
19	1330	1349	1368	1387	1406	1425	1444	1463	1482	1501
20	1400	1420	1440	1460	1480	1500	1520	1540	1560	1580
21	1470	1491	1512	1533	1554	1575	1596	1617	1638	1659
22	1540	1562	1584	1606	1628	1650	1672	1694	1716	1738
23	1610	1633	1656	1679	1702	1725	1748	1771	1794	1817
24	1680	1704	1728	1752	1776	1800	1824	1848	1872	1896
25	1750	1775	1800	1825	1850	1875	1900	1925	1950	1975
26	1820	1846	1872	1898	1924	1950	1976	2002	2028	2054
27	1890	1917	1944	1971	1998	2025	2052	2079	2106	2133
28	1960	1988	2016	2044	2072	2100	2128	2156	2184	2212
29	2030	2059	2088	2117	2146	2175	2204	2233	2262	2291
30	2100	2130	2160	2190	2220	2250	2280	2310	2340	2370
31	2170	2201	2232	2263	2294	2325	2356	2387	2418	2449
32	2240	2272	2304	2336	2368	2400	2432	2464	2496	2528
33	2310	2343	2376	2409	2442	2475	2508	2541	2574	2607
34	2380	2414	2448	2482	2516	2550	2584	2618	2652	2686
35	2450	2485	2520	2555	2590	2625	2660	2695	2730	2765
36	2520	2556	2592	2628	2664	2700	2736	2772	2808	2844

Battalions from which the number

Number to Men of be detached.	70	71	72	73	74	75	76	77	78	79
37	2590	2627	2664	2701	2738	2775	2812	2849	2886	2923
38	2660	2698	2736	2774	2812	2850	2888	2926	2964	3002
39	2730	2769	2808	2847	2886	2925	2964	3003	3042	3081
40	2800	2840	2880	2920	2960	3000	3040	3080	3120	3160
41	2870	2911	2952	2993	3034	3075	3116	5157	3198	3239
42	2940	2982	3024	3066	3108	3150	3192	3234	3276	3318
43	3010	3053	3096	3139	3182	3225	3268	3311	3354	3397
44	3080	3124	3168	3212	3256	3300	3344	3388	3432	3476
45	3150	3195	3240	3285	3330	3375	3420	3465	3510	3555
46	3220	3266	3312	3358	3404	3450	3496	3542	3588	3634
47	3290	3337	3384	3431	3478	3525	3572	3619	3666	3713
48	3360	3408	3456	3504	3552	3600	3648	3696	3744	3792
49	3430	3479	3528	3577	3626	3675	3724	3773	3822	3871
50	3500	3550	3600	3650	3700	3750	3800	3850	3900	3950
51	3570	3621	3672	3723	3774	3825	3876	3927	3978	4029
52	3640	3692	3744	3796	3848	3900	3952	4004	4056	4108
53	3710	3763	3816	3869	3922	3975	4028	4081	4134	4187
54	3780	3834	3888	3942	3996	4050	4104	4158	4212	4266
55	3850	3905	3960	4015	4070	4125	4180	4235	4290	4345
56	3920	3976	4032	4088	4144	4200	4256	4312	4368	4424
57	3990	4047	4104	4161	4218	4275	4332	4389	4446	4503
58	4060	4118	4176	4234	4292	4350	4408	4466	4524	4582
59	4130	4189	4248	4307	4366	4425	4484	4543	4602	4661
60	4200	4260	4320	4380	4440	4500	4560	4620	4680	4740
61	4270	4331	4392	4453	4514	4575	4636	4697	4758	4819
62	4340	4402	4464	4526	4588	4650	4712	4774	4836	4898
63	4410	4473	4536	4599	4662	4725	4788	4851	4914	4977
64	4480	4544	4608	4672	4736	4800	4864	4928	4992	5056
65	4550	4615	4680	4745	4810	4875	4940	5005	5070	5135
66	4620	4686	4752	4818	4884	4950	5016	5082	5148	5214
67	4690	4757	4824	4891	4958	5025	5092	5159	5226	5293
68	4760	4828	4896	4964	5032	5100	5168	5236	5304	5372
69	4830	4899	4968	5037	5106	5175	5244	5313	5382	5451
70	4900	4970	5040	5110	5180	5250	5320	5390	5460	5530
71	4970	5041	5112	5183	5254	5325	5396	5467	5538	5609

in the margin are to be detached.

Number of Men to be detached.	80	81	82	83	84	85	86	87	88	89
2	160	162	164	166	168	170	172	174	176	178
3	240	243	246	249	252	255	258	261	264	267
4	320	324	328	332	336	340	344	348	352	356
5	400	405	410	415	420	425	430	435	440	445
6	480	486	492	498	504	510	516	522	528	534
7	560	567	574	581	588	595	602	609	616	623
8	640	648	656	664	672	680	688	696	704	712
9	720	729	738	747	756	765	774	783	792	801
10	800	810	820	830	840	850	860	870	880	890
11	880	891	902	913	924	935	946	957	968	979
12	960	972	984	996	1008	1020	1032	1044	1056	1068
13	1040	1053	1066	1079	1092	1105	1118	1131	1144	1157
14	1120	1134	1148	1162	1176	1190	1204	1218	1232	1246
15	1200	1215	1230	1245	1260	1275	1290	1305	1320	1335
16	1280	1296	1312	1328	1344	1360	1376	1392	1408	1424
17	1360	1377	1393	1411	1428	1445	1462	1479	1496	1513
18	1440	1458	1476	1494	1512	1530	1548	1566	1584	1602
19	1520	1539	1558	1577	1596	1615	1634	1653	1672	1691
20	1600	1620	1640	1660	1680	1700	1720	1740	1760	1780
21	1680	1701	1722	1743	1764	1785	1806	1827	1848	1869
22	1760	1782	1804	1826	1848	1870	1892	1914	1936	1958
23	1840	1863	1886	1909	1932	1955	1978	2001	2024	2047
24	1920	1944	1968	1992	2016	2040	2064	2088	2112	2136
25	2000	2025	2050	2075	2100	2125	2150	2175	2200	2225
26	2080	2106	2132	2158	2184	2210	2236	2262	2288	2314
27	2160	2187	2214	2241	2268	2295	2322	2349	2376	2403
28	2240	2268	2296	2324	2352	2380	2408	2436	2464	2492
29	2320	2349	2378	2407	2436	2465	2494	2523	2552	2581
30	2400	2430	2460	2490	2520	2550	2580	2619	2640	2670
31	2480	2511	2542	2573	2604	2653	2666	2697	2728	2759
32	2560	2592	2624	2656	2688	2720	2752	2784	2816	2848
33	2640	2673	2706	2739	2772	2805	2838	2871	2904	2937
34	2720	2754	2788	2822	2856	2890	2924	2958	2992	3026
35	2800	2835	2870	2905	2940	2975	3010	3045	3080	3115
36	2880	2916	2952	2988	3024	3060	3096	3132	3158	3204

Battalions from which the number

Number of Men to be detached.	80	81	82	83	84	85	86	87	88	89
37	2960	2997	3034	3071	3108	3145	3182	3219	3256	3293
38	3040	3078	3116	3154	3192	3230	3268	3306	3344	3382
39	3120	3159	3198	3237	3276	3315	3354	3393	3432	3471
40	3200	3240	3280	3320	3360	3400	3440	3480	3520	3560
41	3280	3321	3362	3403	3444	3485	3526	3567	3608	3649
42	3360	3402	3444	3486	3528	3570	3612	3654	3696	3738
43	3440	3483	3526	3569	3612	3655	3698	3741	3784	3827
44	3520	3564	3608	3652	3696	3740	3784	3828	3872	3916
45	3600	3645	3690	3735	3780	3825	3870	3915	3960	4005
46	3680	3726	3772	3818	3864	3910	3956	4002	4048	4094
47	3760	3807	3854	3901	3948	3995	4042	4089	4136	4183
48	3840	3888	3936	3984	4032	4080	4128	4176	4224	4272
49	3920	3969	4018	4067	4116	4165	4214	4263	4312	4361
50	4000	4050	4100	3150	4200	4250	4300	4350	4400	4450
51	4080	4131	4182	4233	4284	4335	4386	4437	4488	4539
52	4160	4212	4264	4316	4368	4420	4472	4524	4576	4628
53	4240	4293	4346	4399	4452	4505	4558	4611	4664	4717
54	4320	4374	4428	4482	4536	4590	4644	4698	4752	4806
55	4400	4455	4510	4565	4620	4675	4730	4785	4840	4895
56	4480	4536	4592	4648	4704	4760	4816	4872	4928	4984
57	4560	4617	4674	4731	4788	4845	4902	4959	5016	5073
58	4640	4698	4756	4814	4872	4930	4988	5046	5104	5162
59	4720	4779	4838	4897	4956	5015	5074	5133	5192	5251
60	4800	4860	4920	4980	5040	5100	5160	5220	5280	5340
61	4880	4941	5002	5063	5124	5185	5246	5307	5368	5429
62	4960	5022	5084	5146	5208	5270	5332	5394	5456	5518
63	5040	5103	5166	5229	5292	5355	5418	5481	5544	5607
64	5120	5184	5248	5312	5376	5440	5504	5568	5632	5696
65	5200	5265	5330	5395	5460	5525	5590	5655	5720	5785
66	5280	5346	5412	5478	5544	5610	5676	5742	5808	5874
67	5360	5427	5494	5561	5628	5695	5762	5829	5896	5963
68	5440	5508	5576	5644	5712	5780	5848	5916	5984	6052
69	5520	5589	5658	5727	5796	5865	5934	6003	6072	6141
70	5600	5670	5740	5810	5880	5950	6020	6090	6160	6230
71	5680	5751	5822	5893	5964	6035	6106	6177	6248	6319

in the margin are to be detached.

Number of Men to be detached	90	91	92	93	94	95	96	97	98	99
2	180	182	184	186	188	190	192	194	196	198
3	270	273	276	279	282	285	288	291	294	297
4	360	364	368	372	376	380	384	388	392	396
5	450	455	460	465	470	475	480	485	490	495
6	540	546	552	558	564	570	576	582	588	594
7	630	637	644	651	698	665	672	679	686	693
8	720	728	736	744	752	760	768	776	784	792
9	810	819	828	837	846	855	864	873	882	891
10	900	910	920	930	940	950	960	970	980	990
11	990	1001	1012	1023	1034	1045	1056	1067	1078	1089
12	1080	1092	1104	1116	1128	1140	1152	1164	1174	1188
13	1170	1183	1196	1209	1222	1235	1248	1261	1274	1287
14	1260	1274	1288	1302	1316	1330	1344	1358	1372	1386
15	1350	1365	1380	1395	1410	1425	1440	1455	1470	1485
16	1440	1456	1472	1488	1504	1520	1536	1552	1568	1584
17	1530	1547	1564	1581	1598	1615	1632	1649	1666	1683
18	1620	1638	1656	1674	1692	1710	1728	1746	1764	1782
19	1710	1729	1748	1767	1786	1805	1824	1843	1862	1881
20	1800	1820	1840	1860	1880	1900	1920	1940	1960	1980
21	1890	1911	1932	1953	1974	1995	2016	2037	2058	2079
22	1980	2002	2024	2046	2068	2090	2112	2134	2156	2178
23	2070	2093	2116	2139	2162	2185	2208	2231	2254	2277
24	2160	2184	2208	2232	2256	2280	2304	2328	2352	2376
25	2250	2275	2300	2325	2550	2375	2400	2425	2450	2475
26	2340	2366	4392	2418	2444	2470	2496	2522	2548	2574
27	2430	2457	2484	2511	2538	2565	2592	2619	2646	2673
28	2520	2548	2576	2604	2632	2660	2688	2716	2744	2772
29	2610	2639	2668	2697	2726	2755	2784	2813	2842	2871
30	2700	2730	2760	2790	2820	2850	2880	2910	2940	2970
31	2790	2821	2852	2883	2914	2945	2976	3007	3038	3069
32	2880	2912	2944	2976	3008	3040	3072	3104	3136	3168
33	2970	3003	3036	3069	3102	3135	3168	3210	3234	3267
34	3060	3094	3128	3162	3196	3230	3264	3298	3332	3366
35	3150	3185	3220	3255	3290	3325	3360	3395	3430	3465
36	3240	3276	3312	3348	3384	3420	3456	3492	3528	3564

Battalions from which the number

Number of Men to be detached.	90	91	92	93	94	95	96	97	98	99
37	3330	3367	3404	3441	3478	3515	3552	3589	3626	3663
38	3420	3458	3496	3534	3572	3610	3648	3686	3724	3762
39	3510	3549	3588	3627	3666	3705	3744	3783	3822	3861
40	3600	3640	3680	3720	3760	3800	3840	3880	3920	3960
41	3690	3731	3772	3813	3854	3895	3936	3977	4018	4059
42	3780	3822	3864	3906	3948	3990	4032	4074	4116	4158
43	3870	3913	3956	3999	4042	4085	4128	4171	4214	4257
44	3960	4004	4048	4092	4136	4180	4224	4268	4312	4356
45	4050	4095	4140	4185	4230	4275	4320	4365	4410	4455
46	4140	4186	4232	4278	4324	4370	4416	4462	4508	4554
47	4230	4277	4324	4371	4418	4465	4512	4559	4606	4653
48	4320	4368	4416	4464	4512	4560	4608	4656	4704	4752
49	4410	4459	4508	4557	4606	4655	4704	4753	4802	4851
50	4500	4550	4600	4650	4700	4750	4800	4850	4900	4950
51	4590	4641	4692	4743	4794	4845	4896	4947	4998	5049
52	4680	4732	4784	4836	4888	4940	4992	5044	5096	5148
53	4770	4823	4876	4929	4982	5035	5088	5141	5194	5247
54	4860	4914	4968	5022	5076	5130	5184	5238	5292	5346
55	4950	5005	5060	5115	5170	5225	5280	5335	5390	5445
56	5040	5096	5152	5208	5264	5320	5376	5432	5488	5544
57	5130	5187	5244	5301	5358	5415	5472	5529	5586	5643
58	5220	5278	5336	5394	5452	5510	5568	5626	5684	5742
59	5310	5369	5428	5487	5546	5605	5664	5723	5782	5841
60	5400	5460	5520	5580	5640	5700	5760	5820	5880	5940
61	5490	5551	5612	5673	5734	5795	5856	5917	5978	6039
62	5580	5642	5704	5766	5828	5890	5952	6014	6076	6138
63	5670	5733	5796	5959	5922	5985	6048	6111	6174	6237
64	5760	5824	5888	5952	6016	6080	6144	6208	6272	6336
65	5850	5915	5980	6045	6110	6175	6240	6305	6370	6435
66	5940	6006	6072	6138	6204	6270	6336	6402	6468	6534
67	6030	6097	6164	6231	6298	6365	6432	6499	6566	6633
68	6120	6188	6256	6324	6392	6460	6528	6596	6664	6732
69	6210	6279	6348	6417	6486	6555	6624	6693	6762	6831
70	6300	6370	6440	6510	6580	6650	6720	6790	6860	6930
71	6390	6461	6532	6603	6674	6745	6816	6887	6958	7029

in the margin are to be detached.

Number of Men to be detached.	100	101	102	103	104	105	106	107	108	109
2	200	202	204	206	208	210	212	214	216	218
3	300	303	306	309	312	315	318	321	324	327
4	400	404	408	412	416	420	424	428	432	436
5	500	505	510	515	520	525	530	535	540	545
6	600	606	612	618	624	630	636	642	648	654
7	700	707	714	721	728	735	742	749	756	763
8	800	808	816	824	832	840	848	856	864	872
9	900	909	918	927	936	945	954	963	972	981
10	1000	1010	1020	1030	1040	1050	1060	1070	1080	1090
11	1100	1111	1122	1133	1144	1155	1166	1177	1188	1199
12	1200	1212	1224	1236	1248	1260	1272	1284	1296	1308
13	1300	1313	1326	1339	1352	1365	1378	1391	1404	1417
14	1400	1414	1428	1442	1456	1470	1484	1498	1512	1526
15	1500	1515	1530	1545	1560	1575	1590	1605	1620	1635
16	1600	1616	1632	1648	1664	1680	1696	1712	1728	1744
17	1700	1717	1734	1751	1768	1785	1802	1819	1836	1853
18	1800	1818	1836	1854	1872	1890	1908	1926	1944	1962
19	1900	1919	1938	1957	1976	1995	2014	2033	2052	2071
20	2000	2020	2040	2060	2080	2100	2120	2140	2160	2180
21	2100	2121	2142	2163	2184	2205	2226	2247	2268	2289
22	2200	2222	2244	2266	2288	2310	2332	2354	2376	2398
23	2300	2323	2346	2369	2392	2415	2438	2461	2484	2507
24	2400	2424	2448	2472	2496	2520	2544	2568	2592	2616
25	2500	2525	2550	2575	2600	2625	2650	2675	2700	2725
26	2600	2626	2652	2678	2704	2730	2756	2782	2808	2834
27	2700	2727	2754	2781	2808	2835	2862	2889	2916	2943
28	2800	2828	2856	2884	2912	2940	2968	2996	3024	3052
29	2900	2929	2958	2987	3016	3045	3074	3103	3132	3161
30	3000	3030	3060	3090	3120	3150	3180	3210	3240	3270
31	3100	3131	3162	3193	3224	3255	3286	3317	3348	3379
32	3200	3232	3264	3296	3328	3360	3392	3424	3456	3488
33	3300	3333	3366	3399	3432	3465	3498	3531	3564	3597
34	3400	3434	3468	3502	3536	3570	3604	3638	3672	3706
35	3500	3535	3570	3605	3640	3675	3710	3745	3780	3815
36	3600	3636	3672	3708	3744	3780	3816	3852	3888	3924

Battalions from which the number

Number of Men to be detached.	100	101	102	103	104	105	106	107	108	109
37	3700	3737	3774	3811	3848	3885	3922	3959	3996	4033
38	3800	3838	3876	3914	3952	3990	4028	4066	4104	4142
39	3900	3939	3978	4017	4056	4095	4134	4173	4212	4251
40	4000	4040	4080	4120	4160	4200	4240	4280	4320	4360
41	4100	4141	4182	4223	4264	4305	4346	4387	4428	4469
42	4200	4242	4284	4326	4368	4410	4452	4494	4536	4578
43	4300	4343	4386	4429	4472	4515	4558	4601	4644	4687
44	4400	4444	4488	4532	4576	4620	4664	4708	4752	4796
45	4500	4545	4590	4635	4680	4725	4770	4815	4860	4905
46	4600	4646	4692	4738	4784	4830	4876	4922	4968	5014
47	4700	4747	4794	4841	4888	4935	4982	5029	5076	5123
48	4800	4848	4896	4944	4992	5040	5088	5136	5184	5232
49	4900	4949	4998	5047	5096	5145	5194	5243	5292	5341
50	5000	5050	5100	5150	5200	5250	5300	5350	5400	5450
51	5100	5151	5202	5253	5304	5355	5406	5457	5508	5559
52	5200	5252	5304	5356	5408	5460	5512	5564	5616	5668
53	5300	5353	5406	5459	5512	5565	5618	5671	5724	5777
54	5400	5454	5508	5562	5616	5670	5724	5778	5832	5886
55	5500	5555	5610	5665	5720	5775	5830	5885	5940	5995
56	5600	5656	5712	5768	5824	5880	5936	5992	6048	6104
57	5700	5757	5814	5871	5928	5985	6042	6099	6156	6213
58	5800	5858	5916	5974	6032	6090	6148	6206	6264	6322
59	5900	5959	6018	6077	6136	6195	6254	6313	6372	6431
60	6000	6060	6120	6180	6240	6300	6360	6420	6480	6540
61	6100	6161	6222	6283	6344	6405	6466	6527	6588	6649
62	6200	6262	6324	6386	6448	6510	6572	6634	6696	6758
63	6300	6363	6426	6489	6552	6615	6678	6741	6804	6867
64	6400	6464	6528	6592	6656	6720	6784	6848	6912	6976
65	6500	6565	6630	6695	6760	6825	6890	6955	7020	7085
66	6600	6666	6732	6798	6864	6930	6996	7062	7128	7194
67	6700	6767	6834	6901	6968	7035	7102	7169	7236	7303
68	6800	6868	6936	7004	7072	7140	7208	7276	7344	7412
69	6900	6969	7038	7107	7176	7245	7314	7383	7452	7521
70	7000	7070	7140	7210	7280	7350	7420	7490	7560	7630
71	7100	7171	7242	7313	7384	7455	7526	7597	7668	7739

CHAP. XX.

Of the CAVALRY.

ARTICLE I.

Directions for the forming of Squadrons, posting the Officers, sending for and returning the Standards, &c.

IS presumed, that the men are taught to ride, and the horses dressed, in order to perform the following Exercise.

The proper arms for a Trooper or Dragoon, are a carbine, or firelock, pistols, and a broad-sword.

The firelock is to be placed in a bucket (which is fixed by Straps to the right side of the saddle, so as to hang below the holster-pipe) and to be fastened about twelve inches above the Lock, by a Strap that comes from the bur, or fore-part of the saddle, the barrel upward, and running between the man's right arm and side.

The sword is to be placed on the man's left thigh, the point something lower than the hilt. As to the placing the pistols, and other accoutrements, it is so generally known, that it will be unnecessary to mention it. If they link with collars (as is customary at present) the end is to be fastened to the right side of the saddle, above the holster-pipe, by a running knot.

When the Regiment is ordered to draw out to exercise, the men are to parade at their Captain's quarters, or place appointed by him, completely armed and accoutred. The Lieutenant, Cornet, and Quarter-Master are to be there likewise at the time appointed; from which place the Captain or Officer commanding the Troop is to march them to the general place of parade, where they are to form in Squadron.

A Regiment confisting of nine Troops is formed into three Squadrons, three Troops in each. The first Squadron is compofed of the Colonel's, firft and fourth Captains Troops. The fecond Squadron, of the Lieutenant-Colonel's, fecond and fifth Captains Troops. The third Squadron, of the Major's, third and fixth Captains Troops.

A Regiment of fix Troops is formed into two Squadrons. The firft Squadron is compofed of the Colonel's, Major's, and fecond Captain's Troops. The fecond Squadron is compofed of the Lieutenant-Colonel's, firft and third Captains Troops.

When the Troops are come to the general place of parade, and form in Squadron, they are to draw up in three Ranks (which is called three deep) and to complete their files. The Officers are to remain at the head of their Troops, the Captain in the center, the Lieutenant on the right, and the Cornet on the left, and the Quarter-Mafter in the rear of the Troop. Care is to be taken, that the Troops are well fized, that is, the talleft men and horfes are to be in the front and rear-ranks, and the loweft in the center.

The Troops that compofe the feveral Squadrons are to draw up in the following manner in Squadron. The eldeft Troop of each Squadron is to be on the right, the fecond on the left, and the youngeft in the center; except the Lieutenant-Colonel's Squadron, which is on the left of the Regiment, and is to draw up the reverfe; the Lieutenant-Colonel's Troop is to be on the left, the next eldeft on the right, and the youngeft in the center.

If a Regiment be compofed of more than two Squadrons, the others are to be drawn up in the fame manner as the firft; but in cafe the Lieutenant-Colonel's Squadron be feparated from the Colonel's, the Lieutenant-Colonel's Troop is then to take the right.

A Regiment confifting of three Squadrons is to be drawn up as follows; the eldeft Squadron on the right, the fecond on the left, and the youngeft in the center. The interval, or diftance between each Squadron, is to be

Chap. XX. *Military Discipline.*

be equal to the ground one Squadron stands on. The distance between the Ranks, when drawn up in Squadron, is to be at *open order*; which is explained in the following article.

The several distances between the Ranks are, *open order, order, close order,* and *close to the croop*

Open order, is the distance between each Rank when drawn up into Squadron, which distance must be equal to half the front of the Squadron.

Order, is the distance the Ranks are to be at, when the Squadrons march, which is equal to a third of the front.

Close order, is the distance the Ranks are to be at, when moving up to the enemy, which distance is, that four men may just wheel round.

Close to the croop, is as close as they can be; in which position they are to charge.

When the Squadrons are formed, the Major or Officer that is to excercise the Regiment, is to order the Officers to take their posts, by saying, *Officers to your posts!* The Officers of each Squadron are to take their posts at the head of their Squadrons by seniority of commission, that is, the eldest Captain on the right, the second on the left, the next eldest on the right, and the next eldest Officer on the left, and so on till the youngest comes in the center; taking care to divide the ground equally between them, so as to cover the front of the Squadron. When the Officers have taken their posts, they are to dress in the same rank, and to be advanced a horse's length before the front rank of men. The Quarter-Masters are to post themselves by seniority in the rear of their respective Squadrons, a horse's length behind the rear rank. The Trumpets or Drums are to draw up on the right of their Squadrons even with the front rank of men, leaving a small interval between them and the Squadron. The Kettle-Drummer is to place himself on the right of the Trumpets or Drums of the Colonel's Squadron. This rule does not extend to the Lieutenant-Colonel's Squadron, when the Colonel's is present, but the Officers are to take their posts from the left to the right, and the Quarter-Masters of that Squadron are to do the same, and the Trumpets or Drums are to place themselves on the left, as the others do on the right.

As soon as the Officers are posted in the manner aforesaid, the Standards are to be sent for in the following manner. The Trumpets or Drums, and Kettle-Drummer, are to be ordered to the center Squadron, where they are to be formed into ranks according to their number, and the Kettle-Drummer advanced before them, all facing outward. This being done, the Major is to order the eldest Cornet of that Squadron to march for the Standards, with a Quarter-Master, and four, five or more files from the center of that Squadron, the Cornet marching at the head of the said detachment, the Trumpets or Drums, and Kettle-Drummer before him, and the Quarter-Master in the rear of the whole. When the Cornet comes to the place where the Standards are lodged (which is always at the Colonel's or commanding Officer's of the Regiment's quarters) he is to form his detachment into a rank entire, facing the house, by saying, *To the right*, or *left* (according as it stands) *form a rank entire! March!* The Kettle-Drummer and Trumpets or Drums are to form into a rank entire with the detachment. When this is done, the Cornet is to order the detachment (the men that are to carry the Standards excepted) to draw their swords (it being the custom of the Cavalry to go for the Standards without sound of Trumpet, beat of Drum, or drawn swords) which being done, and the Standards received, which must be by those of the front rank, he is to form his detachment by these words of command, *To the right* (or left) *form your ranks! March!* At which the Trumpets or Drums are likewise to form as before. The Cornet is then to march back to the Regiment with the Standards, the Trumpets sounding, or Drums beating, a *March*; but, instead of marching along the front of the Regiment, as they do in the Foot, he is to march along the rear, till he comes to the interval, which was made by their marching out, and then he is to wheel and march his men into their former places. As soon as the Standards are come near the

Regiment, the Major is to order the men to draw their swords; which is a ceremony always to be paid the Standards, both in bringing them to, and carrying them from, the Regiment. When the standards are come, the eldest Cornets of the right and left Squadrons are to march with three men from the center of the front rank, along the front, and when they come opposite to the Standards they are to halt, and to order those men that are to carry the Standards to return their swords, and take their several Standards. The Cornets are then to march back with their Standards, taking with them their respective Trumpets or Drums, who are to sound, or beat, a *March*, and as soon as they have got to the center of the intervals between the Squadrons, they are to wheel to the rear, and march till they come opposite to the intervals of the front and center ranks, and then to wheel to their Squadrons, and march between those ranks, till they come to their places, and then to wheel up. The Trumpets or Drums are then to go to their posts, and the Regiment to return their swords.

When the Standards are to be returned, the Major is to order the Cornets of the right and left Squadrons to carry the Standards to the center Squadron, which they are to do in the same manner they brought them from thence, the Trumpets sounding, or Drums beating, a *March*; and when they have delivered them to that Squadron, the Cornets are to return with the men to their Squadrons, marching between the front and center ranks, till they come to their own places, and then wheel up. The Kettle-Drummer and Trumpets, or Drums, are to remain with the Standards, and form their ranks as they did before. This being done, and the Cornets of the other Squadrons returned, the Cornet that brought the Standards is to march back with the same number of files (the Trumpets sounding, or Drums beating, a *March*) and lodge the Standards, drawing up his men in a rank entire as he did when he received them, and then to return the swords, form them into ranks, and march back to the Regiment without sound of Trumpet, or beat of Drum.

Note. Before the Standards are carried to the center Squadron, the Major is to order the Regiment to draw their swords, and as soon as the Standards are gone, to return them.

A Regiment of two Squadrons must send a detachment from the first Squadron for the Standards. All the other ceremonies are to be observed, as before mentioned.

The Standards being brought to their respective Squadrons, the next thing to be done is, the telling off, or dividing the several Squadrons into proper divisions for the Exercise. *First*, each Squadron is to be told off by *files*, then *ranks by fours*; *quarter ranks*; *ranks by three divisions*, and *half ranks*.

Telling off by files. You begin at the right of each Squadron, and say to the right-hand file, *You stand*; To the second, *You move*; the third, *You stand*; the fourth, *You move*; and so on through the Squadrons.

Ranks by fours. You begin at the right of each rank, and say to the first man, *You are the right-hand man of ranks by fours*; to the fourth man, *You are the left-hand man of ranks by fours*; to the fifth man, *You are the right-hand man*, &c. to the eighth man, *You are the left-hand man*, &c. to the ninth, *You are the right*, &c. to the twelfth, *You are the left*, &c. and so on through each Squadron.

Quarter ranks. Each Squadron is to be divided into four equal parts, which are to be called, *First, second, third, and fourth quarter ranks*, beginning at the right, by saying to the right-hand man, *You are the right-hand man of the first quarter rank*; and to the left-hand man of that quarter rank, *You are the left*; and so on to the rest in the same manner.

Ranks by three divisions. Each Squadron is to be divided into three equal parts. The right and left-hand men are to be told in the same manner as the others.

Ranks by two divisions. Each Squadron is to be divided into two equal parts. The right and left-hand men are to be told as above.

Chap. XX. *Military Difcipline.*

As the telling off, or dividing each Squadron into the above-mentioned divifions, will prove tedious by doing it diftinctly through every rank, I will lay down a fhorter method of performing it.

The Major is to order the center and rear ranks to clofe to the croop, at which time the Officers are to advance fo far, that the Major or Adjutant may go with eafe between them and the front rank. By the ranks being clofed, the telling off in the front will ferve for the other ranks. When they are told off, the ranks are to be opened backward to open order, and the Officers, at the fame time, are to rein back to their former diftance.

For the better underftanding the feveral wheelings, it will be proper to give an explanation The circle is divided into four parts. Wheeling to the right or left is a quarter of the circle; to the right or left-about is one half; the circle entire is quite round till you come on the former ground.

Rules for wheeling.

When you wheel to the right, you are to clofe to the right, and look to the left; and when you wheel to the left, you are to clofe to the left, and look to the right. This rule will ferve for all wheeling by ranks; but when you wheel in Squadron, or by divifions, the following rules muft be obferved.

When you wheel by Squadron, or by divifions, the three ranks wheel together, the center and rear ranks wheeling directly in the rear of the front rank, keeping their proper diftance, and each man covering his fileleader.

The firft rank of each Squadron or divifion is to obferve the fame rules as are already given; that is, when you wheel to the right to clofe to the right, and look to the left; and when you wheel to the left, to clofe to the left, and look to the right: but the center and rear ranks are to clofe to the left, when they wheel to the right, that the men may keep oppofite to their file-leaders,

leaders, and those ranks are to move quicker than the front, the circumference they take being larger.

The center and rear ranks are to take particular care, not to close their ranks in wheeling, but to keep directly behind their file-leaders. All wheelings are to be done briskly, but no man to exceed a large trot. The motion of each man is quicker or slower, according to the distance he is from the right or left: thus, when you wheel to the right, each man from the right moves quicker than his right-hand man, the circle that every man wheels being larger, according to the distance he is from the right. When you wheel to the left, the motion of every man is also different, according to the distance he is from the left.

All things being thus disposed, they may proceed to the Exercise; unless they are to perform it before a General Officer; for the reception of whom I will give some directions at the end of this Chapter; as also the manner of passing in review, and Officers saluting.

The Major, or Officer, that is to exercise the Regiment is to place himself opposite to the center of the center Squadron, if the Regiment consists of three Squadrons; but if only of two, he is to place himself opposite to the center of the interval between them.

Those who think the following Exercise too long to be performed at one time (as, no doubt, most people will) may very easily shorten it, by leaving out those things which they do not approve of, which, in my opinion, is a sufficient answer to an objection of that kind.

A R-

ARTICLE II.

The words of command for the manual Exercise, and Evolutions of the Horse, with the explanation, calculated for a Regiment of 3 Squadrons.

I. *Return your Swords!* 3 motions.

1. Bring the sword over the left arm, and enter the point in the scabbard. 2. Thrust it quite up to the hilt. 3. Bring your right hand back to its proper place.

Note, The Officers are to perform these motions with the men.

II. *Files, to the right double! march!*

The 2d, 4th, 6th, and every even file of each Squadron, is to rein back to half the distance between the ranks, then passage a little to the right, to cover their File-leaders.

III. *Make ready your carbines!*

Every man is to swivel and unstrap his carbine, and then buckle the straps, as before.

IV. *Handle your carbines!* 2 motions.

First, extend your arm in a direct line with your shoulder; and, 2. bring it briskly down, and seize the carbine a little above the lock with a full hand, the fingers underneath, and the thumb upon the upper side of the bar.

V. *Sling your carbines!* 4 motions.

1. Pull your carbine out of the bucket, by the same line of direction it was in before, only raising the muzzle a little higher. 2. Turn the butt straight upwards, the barrel from you, and bring the carbine perpendicular before the right breast. 3. Drop it down gently by the inside of the right toe. 4. Bring the right hand
up

up by the inside of the butt, and seize the carbine full on the outside, a little behind the lock.

VI. *Throw back your carbines!* 3 motions.

1. Raise the muzzle of the carbine, as high as the butt, at the same time raising your wrist and elbow square with each other, the muzzle pointing to the front. 2. Throw the carbine cross behind you, with the butt to the near side. 3. Bring your hand and shoulder to their proper posture.

VII. *Make ready your links!*

Unbutton your linking reins, and keep one in each hand.

VIII. *Drop your links!*

Drop the reins on each side the horse's neck.

IX. *Quit your right stirrups!* 5 motions.

1. Raise your right toe, and put your foot back out of the stirrup, and, at the same time, seize the reins with your right hand, a little above the left. 2. Slipping the reins between the forefinger and thumb, raise your right hand to the upper end of the reins, keeping your elbow square. 3. Cast down the reins to the off-side. 4. Raise your bridle hand, under which you are, at the same time, to seize a part of the mane, and twist it into the bridle hand, with which you are to grasp it firm. 5. You are to bring back your right hand to its former position.

X. *Dismount!* 5 motions.

1. Take hold of the right bur of the saddle with the right hand, placing your fingers on the inside, and your thumb on the out. 2. Raise yourself with the right hand above the saddle, and bring your right leg over to the near side of the horse, with an upright body, taking hold, at the same time, of the cantle, or back part of the saddle, with the right hand, looking full to the right

right of the Squadron. 3. Come to the ground with the right foot, facing full to the rear. 4. Quitting the ſtirrup with the left foot, bring it down to the ground, and place it even with the other. 5. Quit the bridle and mane with the left hand, and, at the ſame time, facing to the left-about, take hold of the left cheek of the bridle with the right hand.

Note. The Officers are to perform with the men all the motions of the two foregoing words of command; after which, their ſervants are to take their horſes from the front of the Squadrons.

XI. *Files that doubled, as you were! March!*

The men of the files, who doubled, lead their horſes into their former places; after which the men are to take off their gloves, and ſecure them between their ſword belts and waiſtcoats.

XII. *Link your horſes!*

The men are to face to the right-about, after which every man is to double link, by fixing his own near rein to the off cheek of his next horſe's bridle, and his off rein to the near cheek of his own.

Note. That three men of each rank, which are commonly thoſe of the flanks and center, are to remain with the horſes, to take care of them during the time the reſt are exerciſing; as alſo one Quarter-Maſter of a Squadron. And though thoſe men remained formerly on horſeback, yet it is evident they are more uſeful on foot; but this may be done either way, as the Commanding Officer ſhall think fit. One Trumpet of a Squadron muſt remain on horſeback, to hold the horſes of the others.

XIII. *Face to your proper front!*

All the men, except thoſe that are appointed to remain with their horſes, face to the left-about, in three motions, as deſcribed in the Manual Exerciſe of the Foot; and the Officers at the ſame time, are to divide them themselves equally on the front of their Squadrons, in the same manner as they did on horseback, the Cornets carrying the Standards.

Note. The Cornets are to carry the Standards in the same manner as the Ensigns of Foot do the Colours, by advancing them in marching by the General, or other forms and ceremonies, saluting as they do, and planting them on their right, during the Exercise, &c.

XIV. *Move clear of your horses!* *March!*

Every man is to march three paces from his horse's head, beginning with the left foot, and bringing up his right foot in a line with the other at the last pace, having his carbine ordered, with the butt-end on the right toe.

Note. That in all marching to the front, the men begin with their left feet.

XV. *Rest your carbines!* 3 motions.

This is done as in the Foot Exercise, from the position of ordered arms.

XVI. *Unspring your carbines!* 3 motions.

1. Take hold of the swivel with your right hand, placing your thumb on the spring, and pressing it down at the same time, take it out of the ring, and keep your elbow square. 2. Throw off the swivel to the right side. 3. Bring the right hand up to its former place.

XVII. *Shoulder!* 4 motions.

As in the Foot Exercise.

Note. At this word of Command, the Officers are to draw their swords.

XVIII. *Squadrons, face to the right!*

The Officers and men face to the right in two motions, as directed in the Foot Excercise.

XIX. *March!*

They all begin their march at the same time with the left foot, moving very flow; and when the left of each Squadron is two paces beyond the right of their horses, the following word of command is to be given.

XX. *Halt!*

They all face to the left in two motions, as described in the Exercise of the Foot, and immediately straighten their ranks.

XXI. *Center and rear ranks, close to your proper distance! March!*

The center and rear ranks of each Squadron march forward, till they come within six paces of each other, and then halt. After this, the files are to be compleated, the ranks dressed, and the Battalion told off. Then the Major proceeds.

Squadrons, have a care to march forward!

At this the Cornets advance the Standards.

XXII. *March!*

The three Squadrons march straight forward, keeping in a line, taking care not to open or close their ranks and files, and covering their file leaders; and when they are advanced to a proper distance from the horses, the following word of command is to be given.

XXIII. *Halt!*

At this they all stand, and immediately straighten their ranks and files. The Cornets plant their Standards on the right, as the Ensigns do the Colours.

XXIV. *To the left open your files!*

All face to the left in two motions,

XXV. *March!*

The left hand file of each squadron begins at the same time with the right feet, and marches very slow. The rest of the files are to move as soon as those before them are at a proper distance; and when the whole are opened, and the three Squadrons joined, the next word of command is to be given.

XXVI. *Halt!*

They are to face to the right in two motions.

XXVII. *Standards take your posts in the center!*

The Cornets who carry the Standards of the right and left Squadrons, face to the right and left inwards; that is, the Cornet on the right, faces to the left; and the Cornet on the left, faces to the right, and then wait for the following word.

XXVIII. *March!*

The two Cornets with the Standards, march along the front of the rank of Officers to the center; and as soon as they have got on the right and left of the Cornet who carries the Standard of the center Squadron, they are to stand facing one another, till the next command.

XXIX. *Halt!*

The two Cornets with the Standards face, in two motions, to their proper front, and dress in a line with the Cornet, who carries the center Standard.

All the other Officers remain in their former posts, as when drawn up in Squadron; only they are all to divide the ground equally between them, and dress in the same line.

XXX. *Officers, take your posts of Exercise!*

The Officers face to the right-about in three motions. Half the Quarter-Masters on the right, face to the right, and

and the other half face to the left, in two motions, and all remain in this pofition till the next word of command.

Note, The Colonel, or Officer commanding the Regiment in his abfence, is not to face with the Officers, but remain facing to the front; together with the three divifions of Trumpeters on the flanks and center.

XXXI. *March!*

The Colonel, or Commanding Officer, marches ftraight forward, and places himfelf by the General during the Exercife; the other Officers march through the ranks, beginning with the left feet, and keeping in a line, till they get nine paces beyond the rear rank, and then ftand; but the Lieutenant-Colonel, unlefs he commands the Regiment, is to march two paces beyond the Officers.

The Quarter-Mafters are to march pretty quick, and place themfelves on the right and left of each rank. The Trumpets are to march ftraight forward, and place themfelves in the rear of the Major, or Officer that exercifes the Regiment.

XXXII. *Halt!*

The Lieutenant-Colonel, and the reft of the Officers in the rear, face to the left-about.

Take care!

Note, The Foot Exercife being already explained, it will be unneceffary to infert it here; I therefore refer you to the Account of it, both for the manner of performing each motion, and the proper time between them.

    XXXIII. *Reft your carbines!* 3 motions.
    XXXIV. *Order your carbines!* 3 motions.
    XXXV. *Ground your carbines!* 4 motions.
    XXXVI. *Take up your carbines!* 4 motions.
    XXXVII. *Reft your carbines!* 3 motions.
    XXXVIII. *Club your carbines!* 3 motions.
    XXXIX. *Reft your carbines!* 3 motions.
    XL. *Secure your carbines!* 3 motions.

XLI. *Shoulder your carbines!* 3 motions.
XLII. *Officers, take your posts in the front!*
XLIII. *March!*
XLIV. *Present your arms!* 3 motions.
XLV. *To the right!* 3 motions.
XLVI. *To the right!* 3 motions.
XLVII. *To the right-about!* 3 motions.
XLVIII. *To the left!* 3 motions.
XLIX. *To the left!* 3 motions.
L. *To the left-about!* 3 motions.
LI. *Shoulder your carbines!* 2 motions.
LII. *Rest upon your carbines!* 5 motions.

Note, after the performing of this command, the men are to put on their gloves.

LIII. *Shoulder your carbines!* 5 motions.

LIV. *Officers take your posts in the front!*

At this word of command, the Officers in the rear are to be ready to move up to their former posts, as soon as the next is given, and the Cornets are to advance the Standards. The Quarter-Masters on the right, are to face to the left; and those on the left are to face to the right.

LV. *March!*

The Lieutenant-Colonel, with the commissioned Officers, march straight forwards, and place themselves on the front of the Regiment, as before; and the Quarter-Masters march, and post themselves in the rear: The Colonel, or Commanding-Officer, and the Trumpeters, remain in the front, till the following word of command.

LVI. *Standards move to your respective Squadrons!*

The Cornet who carries the Standard on the right, faces to the right; and the Cornet who carries the left Stan-

Standard, faces to the left, and both remain so till the following word of command.

LVII. *March!*

The two Cornets march along the front of the rank of Officers, till they come opposite to the center of their respective Squadrons, and then stand, facing towards the flanks of the Regiment.

The Trumpeters march at the same time, and post themselves on the flanks and center, as before.

LVIII. *Halt!*

The two Cornets, with the right and left Standards, face to their proper front, and immediately dress in a line with the Officers.

LIX. *To the right, close your files!*

They all face to the right.

LX. *March!*

They all begin at the same time with the left feet, marching very slow; and when the right hand file of each Squadron comes to the ground they stood on before they opened their files, they are to stand, and the rest are to march on, and close to them; by which each Squadron will be opposite to the interval on the right of their horses.

LXI. *Halt!*

They all face to the left; and the Trumpeters of the center Squadron post themselves on the right of it, dressing in a line with the front rank.

LXII. *Squadrons, face to the right-about!*

They all face to the right-about.

LXIII. *Ranks, take your open distance! March!*

The center and rear ranks of each Squadron march towards the rear, beginning with their left feet, and moving very slow; and as soon as each rank has got to its open diſtance (which diſtance muſt be equal to half the front of the Squadron) they are to ſtand.

LXIV. *Squadrons, march!*

They all begin with their left feet, and march very ſlow, till each rank comes within a pace of the right of their horſes, and then ſtand facing to the rear.

LXV. *Squadrons, face to the right!*

They all face to the right, in two motions.

LXVI. *March!*

They all begin at the ſame time with the right feet, and march in a ſtraight line, till every man comes oppoſite to his own horſe, and then ſtand, facing to the left of the Squadrons.

LXVII. *Halt!*

At this word of command, they all face to their horſes.

LXVIII. *Unlink your horſes!*

Every man is to ſtep up to the off ſide of his horſe, at the ſame time ſlipping his carbine down from his ſhoulder with the left hand, reſting the cock upon the two laſt fingers, the muzzle upwards, and ſo unlink the horſes.

Note. On the giving of the above word of command, the Officers are to return their ſwords, and their ſervants bring their horſes to the front of the Squadrons.

LXIX. *Files, to the right double! March!*

Every even file reins back, as in explanation 2.

LXX. *Return your carbines!*

Every man is to make a large ſtep with his left foot towards the off ſide of his horſe, and paſſing the muzzle of his carbine through the ſtrap, place the butt in the bucket, and then bring his left foot to its former place.

LXXI.

LXXI. *Fix your links!*

Button the linking reins as before, and take hold of them with the left hand, near the off-cheek, and with the right hand between the buttons.

LXXII. *Throw over your links!*

Throw the linking reins with the right hand over your horse's neck; then with the same hand take hold of the off-cheek of the bridle.

LXXIII. *Face to your proper front!*

Change your right hand from the off, to the near side of the bridle, stepping with the right foot at the same time towards the horse's near shoulder, and, after a short pause, face to the left-about, to your proper front.

LXXIV. *Make ready to mount!* 7 motions.

1. Fall back with the right foot, quitting at the same time the cheek of the bridle; take hold of the end of the reins with the right hand, and put the last fingers of the left between the reins. 2. Raise your reins upright with the right hand, till you can fix them conveniently in the left hand, so as to feel your horse's mouth. 3. Throw the reins with the right hand to the off-side of the horse. 4. Seize a part of your horse's mane with your right hand. 5. Place it in your bridle hand, grasping both the mane and bridle firm with that hand. 6. Take hold of the stirrup with your right hand. 7. Put your left foot in the stirrup.

LXXV. *Mount!* 6 motions.

1. Take hold of the off-side of the cantle, or hind part of the saddle, with the right hand. 2. Raise yourself upright in the stirrup, bringing both feet even together, and looking full to the right of the Squadron.

3. Bring your right leg with a straight body over the horse, and place yourself in the saddle. 4. Seize your rein with the fore finger and thumb of the right-hand, just above the left. 5. Slip your right hand up to the end of the rein, with the elbows square, till you fix your bridle hand in its due position. 6. Throw the reins to the off side of your horse, with the right hand, letting it fall down by your right side, and putting your right foot into the stirrup at the same time.

Note, The Officers are to perform with the men all the motions directed in the two foregoing words of command, as also those for dismounting, as is shewn in Explanation 9, 10.

LXXVI. *Files that doubled, as you were ! March !*

The files that reined back, passage a little to the left, to bring them opposite to their former places, and then move into them pretty briskly, but all at a time.

The Cornets with the Standards are to rein back into the front rank, and the Kettle-Drummer and Trumpets, to post themselves as before.

As this Exercise is calculated for the dismounting and forming into Battalion a Regiment consisting of three Squadrons, it will be proper to give some Directions, how a Regiment of two Squadrons is to perform it.

ARTICLE III.

The words of command for the Exercise of a Regiment of two Squadrons, according to the foregoing directions.

Note, The number to the first word of command, and that which follows the 76th, are omitted, that the other numbers may the better agree with the foregoing explanation.

Squadrons, take your double distance ! March !

This relating only to a Regiment of two Squadrons, the explanation is as follows.

The

Chap. XX. *Military Discipline.* 369

The Squadrons paſſage to the right and left outwards, ſo as to leave an interval of double diſtance.

I. *Return your ſwords!*

From this word of command, to number XVIII, the ſame as in the preceding Exerciſe.

XVIII. *Squadrons, face to the right and left inwards!*

The Officers and men of the right Squadron face to the left, and thoſe of the left Squadron face to the right.

XIX. *March!*

They all begin at the ſame time with the feet they faced on; and when the two Squadrons join in the center of the interval, they ſtand facing one another till the following command.

XX. *Halt!*

They are all to face to their proper front, thoſe of the right Squadron facing to the right, and thoſe of the left Squadron, to the left.

From this word of command, to number LXV, the ſame as above.

LXV. *Squadrons, face to the right and left outwards!*

Thoſe of the firſt Squadron face to the left, and thoſe of the ſecond Squadron face to the right.

LXVI. *March!*

They all begin at the ſame time with the feet next the front, and march in a ſtraight line, till every man comes oppoſite to his own horſe, and then ſtand, facing to the right and left outwards.

LXVII. *Halt!*

At this word of command they all face to their horses.

From this word of command, to number LXXVI, inclusive, the same as in the foregoing Exercise.

Squadrons, close to your proper distance! March!

The Squadrons are to close by passage to the right and left inwards to their usual distance.

Note. The Evolutions are the same as explained in the following Article.

EVOLUTIONS.

ARTICLE IV.

Squadrons, take care to perform your Evolutions!

LXXVII. *Squadrons, from the center open your files!*

The right and left hand file of each Squadron passage to the right and left at the same time, keeping an equal, but slow pace; and as soon as they have got to such a distance from the files which stand next them, that a horse can come between them, those files are to move the same way, and so on, till the two files in the center are opened.

The men of the front rank are to take particular care in opening, to keep the above distance; and those of the center and rear ranks are to keep directly in a line with their file-leaders; and the whole to take care, that they don't advance, or rein back, but passage in a straight line to the right and left. As soon as the two center files are opened, the Major proceeds to the next word of command.

LXXVIII. *Halt!*

At this command the men are to stop their horses, and straighten their ranks and files.

LXXIX.

LXXIX. *Clofe your files to the center! March!*

All the files are to move at the fame time, and clofe by paffage to the center of their refpective Squadrons; but, in order to prevent the horfes from treading on one another, they fhould not clofe till the files on the right and left ftand.

Note, The Officers are to open and clofe with the men, taking care to keep their proper diftances, by each moving an equal pace, and in a direct line with the file in his rear.

LXXX. *Officers take your pofts of Exercife! March!*

The commiffioned Officers march ftraight forwards, the Cornets carrying the Standards till they pafs the Major, or Officer who exercifes the Regiment, 8 or 10 paces, and then ftand, taking care both then, and in marching to the front, that their rank be even.

LXXXI. *Halt!*

The Officers of the right Squadron are to face to the left-about; the Officers of the left Squadron to the right-about, and thofe of the center Squadron to the right and left-about inwards; that is, half the Officers on the right to face to the left-about, and the half on the left to the right-about, in which pofture they are to remain till ordered to their former pofts.

The Kettle-Drummer and Trumpets are to march at the fame time with the Officers; and as foon as the word *Halt!* is given, they are to draw up in the rear of them, where they are to remain.

LXXXII. *Half-ranks to the right double your files! March!*

The left half-ranks of each Squadron rein back to the center of the interval between the ranks, and then paffage to the right in a ftraight line, till the right file comes oppofite to the right hand file of the half-ranks which stand, and then halt, and cover their file-leaders, and straighten their ranks.

This movement must be done together, that all may begin, and finish at the same time. In the passage they are not to open their files, but to keep as close as they can without treading on one another.

LXXXIII. *Half-ranks that doubled, as you were! March!*

The half-ranks that doubled, passage very flow in a direct line to the left, till the right file comes a little to the left of the left hand file of the half-ranks, who stood; after which they are to move up to their former places. Observe further as is directed in the foregoing article.

LXXXIV. *Half-ranks to the left double your files! March!*

The half-ranks on the right of each Squadron rein back, and passage to the left, till the left hand file comes opposite to the left of those that stand, and then halt, taking care to cover their file-leaders, and straighten their ranks.

LXXXV. *Half-ranks that doubled as you were! March!*

The half-ranks that doubled, passage to the right, till the left hand file comes to the right of those that stood, and then move into their places.

LXXXVI. *Rear ranks to the right double your front! March!*

The rear rank of each Squadron divides into two equal parts, and wheels to the right, which forms them into one division of two ranks each, facing to the right: As soon as they have made this wheel, the first rank of each division wheels again to the left, and marches forward, and joins on the right of the front rank; and the
second

Chap. XX.   *Military Difcipline.*   373

fecond rank of each divifion, as foon as the front ranks begin their fecond wheel, marches ftraight forward to the right flank of their Squadrons, and then wheel to the left upon the fame ground the other did, and marches up, and forms on the right of the center rank: Both ranks of each divifion are to march, and wheel in fuch a manner, that they form on the right of their refpective Squadrons at the fame time.

LXXXVII. *Rear ranks that doubled, as you were!*
March!

The rear ranks that doubled, wheel to the right-about, and march directly to the rear, till they come even with the ground they ftood on before, and then wheel to the right, and march towards the left of their Squadrons, till the right hand man of each rank comes oppofite his file-leader, and then they are all to wheel to the right, and form in a rank as before.

LXXXVIII. *Rear ranks to the left double your front!*
March!

The rear rank of each Squadron wheels by two divifions to the left, and forms on the left of their refpective Squadrons, as they did before on the right.

LXXXIX. *Rear ranks that doubled, as you were!*
March!

The rear ranks that doubled, wheel to the left-about, and form on their former ground, as in *Explanation* 87; with this Difference only, that all their wheelings are to the left, as the others were to the right.

XC. *Rear ranks to the right and left, double your front!*
March!

The rear rank of each Squadron divides into four equal parts, or divifions; the two on the right wheel to the right, and the two on the left wheel to the left, which forms the rear rank of each Squadron into two

divifions

divisions of two ranks each, facing to the right and left outwards: After they are thus formed, they are to wheel up to the front, and march, and join on the right and left of the front and center ranks of their respective Squadrons.

XCI. *Rear ranks that doubled, as you were! March!*

The rear ranks that doubled, wheel to the right and left-about, and march to the rear, till they come even with their former ground; then wheel inwards, and march till the flank-men come opposite to their file-leaders; after which they all wheel up, and form a rank entire in the rear of their respective Squadrons.

XCII. *By two divisions wheel to the right and left-about outwards! March!*

Each rank divides in the center, and wheels to the right and left-about outward, which forms each Squadron into two divisions, with an interval between them, facing to the rear.

XCIII. *Wheel to the right and left-about to your proper front! March!*

This wheel brings them into Squadrons, facing to their proper front as before.

XCIV. *Center and rear ranks, move forward to order! March!*

The center and rear ranks of each Squadron move up to order; which distance between the ranks must be equal to a third of the front of the Squadron.

XCV. *By three divisions wheel to the right! March!*

Each rank wheels by three divisions to the right a quarter of the circle.

XCVI.

XCVI. *To the right! March!*

They are to wheel in the same manner another quarter of the circle to the right.

XCVII. *To the right about! March!*

The same divisions are to wheel one half of the circle to the right, which brings the Squadrons to their proper front.

XCVIII. *By three divisions wheel to the left! March!*

Each rank wheels by three divisions a quarter of the circle to the left.

XCIX. *To the left! March!*

They are to wheel another quarter of the circle to the left.

C. *To the left-about! March!*

The same divisions are to wheel half the circle to the left, which brings them again to their proper front.

CI. *Center and rear ranks, move forward to close order! March!*

The center and rear ranks of each Squadron move up so close, that only four horses can just wheel round between the ranks.

CII. *By fours wheel to the right-about! March!*

The ranks being told off by fours, they are to wheel by fours to the right-about; for the performing of which, the right hand man of each rank of fours, is to keep his horse's fore-feet, as much as possible, on the same ground he stood upon; and as he finds the other three men come about, he is to throw the flank of his horse to the left.

CIII. *To the right-about! March!*

The fore-said ranks by fours wheel again to the right-about, which brings them to their proper front.

CIV. *By fours wheel to the left-about! March!*

The same ranks by fours wheel to the left-about, as they did before to the right-about, with this difference only, that the left hand man of each rank of fours is to throw his horse's flank to the right, as the other flank-man did to the left, and to keep his horse's fore-feet on the same ground he stands on, as much as possible.

CV. *To the left-about! March!*

This wheel brings them to their proper front, and must be performed in the same manner, as above directed.

CVI. *Officers take your posts at the head of your Squadrons! March!*

The Officers march in a direct line to the head of their Squadrons; and as soon as they come within half a horse's length of the front rank, they are to stand: The Kettle-drummer and Trumpets are to march at the same time to their former posts on the flanks of their Squadrons, and then stand.

CVII. *Halt!*

At this word of command, the Officers of the right Squadron face to the right-about, and those of the left Squadron face to the left-about: The Officers of the center Squadrons face to the right and left-about inwards, thus; half on the right, to the right-about; and half on the left, to the left-about.

The Trumpets on the right and center Squadrons, face to the right-about, and those of the left Squadron face to the left-about, their posts being on the left of that Squadron.

As soon as the Officers have faced to their proper front, the Cornets, who carry the Standards, are to rein back into the front rank, placing themselves between the two center-men.

CVIII. *Squadrons wheel to the right! March!*

In wheeling by Squadron, the three ranks wheel together, the center and rear ranks wheeling directly to the rear of the front ranks, keeping their proper distance, and each man covering his file-leader.

The Officers wheel on the head of their Squadrons, and the Quarter-masters in the rear. This wheel is a quarter of the circle to the right.

CIX. *Wheel to the right! March!*

The Squadrons are to wheel another quarter of the circle to the right, as before directed.

CX. *Wheel to the right-about! March!*

The Squadrons are to wheel in the same manner, half the circle to the right, which brings them to their proper front.

CXI. *Squadrons wheel to the left! March!*

The Squadrons wheel a quarter of the circle to the left.

CXII. *Wheel to the left! March!*

They wheel another quarter of the circle to the left.

CXIII. *Wheel to the left-about! March!*

This wheel brings the Squadrons to their proper front, and finishes the Evolutions of the Horse.

CXIV. *Center and rear ranks, rein back to your open order! March!*

The center and rear ranks of each Squadron rein back to the ground they first stood on, leaving the distance between

between each rank equal to half the front of the Squadron, that being the true distance of open order, and at which the ranks are always to be, when the Squadrons are drawn up for Exercise.

CXV. *Draw your swords*

Bring your right hand briskly over your left arm, which you must press close to your left side, and seize the handle of your Sword with a full hand: tell 1, 2, and draw your sword quite out of your scabbard, by raising up your right hand as high as your arm will permit, keeping the point downwards: then tell 1, 2, and bring your right hand briskly to your right side, placing the inside of the hilt on the outside of your right thigh, the wrist bending a little out, raising the point pretty high, and running in a line with the off-ear of the horse, with the edge from you.

The Officers are to perform these motions with the men.

CHAP.

CHAP. XXI.

Exercife of the Dragoons.

With the neceffary alterations explained in the points where they differ from the horfe and foot; all the reft of the exercife and motions being the fame with thofe of the horfe and foot, as explained by them.

Words of command.

RETURN your fwords! 3 motions.

II. *Files, to the right double!*

III. *March!*

The men pull off their gloves.

IV. *Make ready your firelocks!* 4 motions.

1. Raife your right hand as high as your fhoulder. Tell 1, 2, and feize your firelock brifkly about eight inches above the lock. Tell 1, 2, and clear it off the bucket with a lively motion. Tell 1, 2, and extend your arm downwards.

V. *Advance your firelocks!* 4 motions.

1. Pull up your firelock, keeping the butt forwards, and, at the fame time, feize it with your left hand a little above the lock. Tell 1, 2, and bring your firelock before you with the left hand, and, at the fame time, feize it with the right hand below the lock, keeping the firelock ftraight up and down, almoft at a recover. Tell 1, 2, and extend both arms brifkly. Tell 1, 2, and, quitting the firelock with your left hand,

hand, bring down the butt on your right thigh, the muzzle sloping a little forwards.

VI. *Poise your firelocks!* 1 motion.

Bring up your firelock to a poise, turning the lock outwards.

VII. *Handle your slings.* 2 motions.

Seize your sling with the left hand, four inches above the lock. Tell 1, 2, and extend your sling to the left, keeping it in a line with the firelock.

VIII. *Sling your firelocks!* 3 motions.

Bring the sling with the left hand opposite to the right shoulder, and the firelock with the left hand opposite to the left shoulder, by crossing of both hands; at the same time, bring the left hand within the right, keeping the muzzle directly up, the barrel to the left, and the right hand just under the left elbow. Tell 1, 2, bend the firelock back, and bring the sling over your head, placing it just above your right shoulder, and the firelock opposite to the point of the left. Tell 1, 2, draw the sling with your left hand, and let go the firelock with the right at the same time, that it may hang by the sling on the right shoulder, the muzzle upwards, and dropping both hands down by your sides at the same time.

IX. *Make ready your links!*
X. *Quit your right stirrup!* 5 motions.
XI. *Dismount!*
XII. *To the left!*

Note. The Officers perform the motions of the above three words of command, and their servants take away their horses.

XIII. *Files that doubled, as you were!*
XIV. *March!*

XV. *To the right-about!*
XVI. *Link your horses!*
XVII. *To the left-about!*
XVIII. *Move clear of your horses!*
XIX. *March!*
XX. *Halt!*
XXI. *Handle your flings!* 3 motions.

1. Seize the fling with both hands at the fame time, taking hold of it with the right hand about the middle, and as low as you can reach, without bending your body, with the left. Tell 1, 2, and, with the left hand, bring the butt forward, flipping your left elbow under the firelock, by bringing of it between the firelock and the fling; take hold of the firelock at the fame time with the left hand, letting the ftock lie between the thumb and fore-finger, the butt-end pointing a little to the left, with the barrel upwards. Tell 1, 2, bring the firelock to lie on the left fhoulder, and the fling on the right, the barrel upwards, and the butt-end pointing directly to the front, keeping the firelock to a true level.

XXII. *Poife your firelocks!* 3 motions.

1. Bring the fling over your head with the right hand, and the firelock ftraight before you with the left, the muzzle upright, and the barrel to the front. Tell 1, 2, and caft the fling brifkly with your right hand towards the left, between the firelock and your body, turning the lock outwards, and at the fame time feize it with your right hand below the cock, the thumb upwards. Tell 1, 2, and thruft the firelock from you to your poife, letting the left hand fall down to your fide.

XXIII. *Shoulder!*

XXIV. *Squadrons, face to the right and left inwards!*

The right hand Squadron faces to the left, the left hand Squadron to the right, and half the center Squadron

Squadron to the right, and half to the left; and, in marching, joins the other two Squadrons in the right and left intervals.

XXV. *March!*
XXVI. *Halt!*
XXVII. *Center and rear ranks, close to your proper distance!*
XXVIII. *March!*

The center and rear ranks of each Squadron march forward, till they come within six paces of each other, and then halt. After this the files are to be completed, the ranks dressed, and the Battalion told off. Then the Major proceeds.

XXIX. *Squadrons, have a care to march forward!*

At this the Cornets advance the standards.

XXX. *March!*

The three Squadrons march straight forward, keeping in a line, taking care not to open or close their ranks and files, and covering their file-leaders; and when they are advanced to a proper distance from their horses, the following word of command is to be given:

XXXI. *Halt!*

At this they all stand, and immediately straighten their ranks and files. The Cornets plant their standards on their right, as the Ensigns do the Colours.

XXXII. *Squadrons, from the right and left face inwards, and form the Battalion!*
XXXIII. *March!*
XXXIV. *Halt!*
XXXV. *Officers, take your posts in Battalion!*

Until this word of command is given the Officers are to remain with their respective Squadrons; but then they are to take their posts on the head of the Battalion by seniority of commission, as they do in the Foot, the

eldest

eldeſt on the right, the next on the left, and ſo on till the Cornets with the Standards come in the center

They are then to divide the ground equally between them, and to dreſs in one rank.

The Quarter-Maſters are to take their poſts in the rear, in the ſame manner.

The Drums are to be divided on the right and left.

XXXVI. *Rear half files, to the right form ſix deep!*

The rear half files face to the right, at two motions.

XXXVII. *March!*

They ſtep off with the right feet, and make eighteen paces to the rear, halting with the left foot foremoſt at the laſt.

XXXVIII. *Halt!*

They come to the left-about in three motions, and inſtantly cover their leaders.

XXXIX. *Officers, recover your arms!*

All the Officers and Quarter-Maſters recover their firelocks in two motions; and the Cornets advance the Standards.

XL. *To the right-about!*

The Officers face to the right-about in three motions. Half the Quarter-Maſters on the right, face to the right, and the other half face to the left, in two motions, and all remain in this poſition till the next word of command.

Note. The Colonel, or Officer commanding the Regiment in his abſence, is not to face with the Officers, but remain facing to the front.

XLI. *March!*

The Colonel, or Commanding Officer, marches straight forward, and places himself by the General during the Exercise; the other Officers march through the ranks, beginning with the left feet, and keeping in a line, till they get nine paces beyond the rear rank, and then stand; but the Lieutenant-Colonel is to march two paces beyond the Officers.

The Quarter-Masters are to march pretty quick, and place themselves on the right and left of each rank. The Drums are to march straight forward, and place themselves in the rear of the Major, or Officer that exercises the Regiment.

The *Troop* beats, while the Officers are thus marching to their posts of exercise.

XLII. *Halt!*

The Lieutenant-Colonel, and the rest of the Officers in the rear, face to the left-about.

XLIII. *Order your arms!*

The Officers order their firelocks in three motions, and remain in that position, the Cornets planting the Standards on the right.

Take care to perform the Manual Exercise!

Note, The Foot Exercise being already explained, it will be unnecessary to insert it here; I therefore refer you to the account of it, both for the manner of performing each motion, and the proper time between them.

44. *Rest your firelocks!* 3 motions.
45. *Order your firelocks!* 3 motions.
46. *Ground your firelocks!* 4 motions.
47. *Take up your firelocks!* 4 motions.
48. *Rest your firelocks!* 3 motions.
49. *Club your firelocks!* 3 motions.
50. *Rest your firelocks!* 3 motions.

Chap. XXI. *Military Discipline,*

51. *Secure your firelocks!* 3 motions.
52. *Shoulder your firelocks!* 3 motions.
53. *Poise your firelocks!* 2 motions.
54. *Rest upon your arms!* 3 motions.
55. *Draw your bayonets!* 2 motions
56. *Fix your bayonets!* 3 motions.
57. *Poise your bayonets!* 3 motions.
58. *Shoulder!* 2 motions.
59. *Present your arms!* 3 motions.
60. *To the right!* 3 motions.
61. *To the right!* 3 motions.
62. *To the right-about!* 3 motions.
63. *To the left!* 3 motions.
64. *To the left!* 3 motions.
65. *To the left-about!* 3 motions.
66. *Charge your bayonets!* 1 motion.
67. *Rest your bayonets on the left arm!* 3 motions.
68. *Rest your bayonets!* 3 motions.
69. *Shoulder!* 2 motions.
70. *Rear half files, to the left double your front! March!* } 18 motions.
71. *Prime and load!* 21 motions.
72. *As front rank, make ready!* 3 motions.
73. *Present!* 1 motion.
74. *Fire!* 19 motions
75. *As center rank, make ready!* 3 motions.
76. *Present!* 1 motion.
77. *Fire!* 19 motions.
78. *As rear rank, make ready!* 3 motions.
79. *Present!* 1 motion.
80. *Fire!* 19 motions.
81. *Rear ranks, close to the front! March!* 10 motions.
82. *Make ready!* 3 motions.
83. *Present!* 1 motion.
84. *Fire!* 1 motion.
85. *Charge your bayonets!* 1 motion
86. *Recover your arms!* 1 motion.
87. *Rear ranks to your proper distance!* 3 motions.
88. *March!* 10 motions.

89. *Halt!* 3 motions.
90. *Shut your pans!* 6 motions.
91. *Clean your bayonets!* 2 motions.
92. *Shoulder!* 2 motions
93. *Rear half files, as you were!* 2 motions.
94. *March!* 18 motions.
95. *Halt!* 3 motions
96. *Front half files, to the right double your rear!* } 3 motions.
97. *March!* 18 motions.
98. *Halt!* 2 motions.
99. *Front half files, as you were! March!* 18 motions.
100. *Officers, recover your arms!* 2 motions.
101. *March!* The Troop beats.
102. *Halt!* 2 motions.
103. *Rear half files, to the left double your front! March!* } 18 motions.
104. *Left hand divisions, by the side step double your files to the right!* } 4 motions.
105. *March!*
106. *Halt!* 1 motion.
107. *Divisions, to the left as you were!* 1 motion.
108. *March!*
109. *Halt!* 4 motions.
110. *Right hand divisions, by the side step double your files to the left!* } 4 motions.
111. *March!*
112. *Halt!* 1 motion
113. *Divisions, to the right as you were!* 1 motion.
114. *March!*
115. *Halt!* 4 motions.
116. *Rear ranks, close to the front!*
117. *March!* 10 motions.
118. *Upon the center, wheel to the right-about!* 3 motions.
119. *March!*
120. *Halt!* 3 motions.
121. *Upon the center, wheel to the left-about!* 3 motions.
122. *March!*
123. *Halt!* 3 motions.

124. *Rear ranks, gain your former distance!* 3 motions.
125. *March!* 10 motions.
126. *Halt!* 3 motions
127. *Poise your firelocks!* 2 motions.
128. *Rest on your arms!* 3 motions.
129. *Unfix your bayonets!* 3 motions.
130. *Return your bayonets!* 2 motions.
131. *Poise your firelocks!* 3 motions.
132. *Shoulder!* 2 motions.
133. *Officers, to your Squadrons!*

The commissioned Officers and Quarter-Masters, face to the right and left, as the Squadrons they belong to are posted, and remain so, till the following word of command. The Cornets are to advance the standards.

CXXXIV. *March!*

The commissioned Officers and Quarter-Masters march to their several Squadrons, and post themselves on the front and rear, as before, by seniority.

The Drummers also march to their respective Squadrons; those of the right Squadron post themselves on the right of it, in a line with the front rank; those of the left Squadron, on the left; and those of the center Squadron, are to post themselves in the rear of the Cornet with the Standard.

135. *Halt!*
136. *From the center face outwards!*
137. *March!*
138. *Halt!*
139. *Dragoons, face to the right-about!*
140. *March!*
141. *Halt!*
142. *To the left-about!*
143. *Dragoons, face to your horses!*
144. *March!*
145. *Halt!*
146. *Poise your firelocks!*
147. *Handle your slings!*

148. *Sling your firelocks!*
149. *To the right-about!*
150. *Move to your Horses!*
151. *March!*
152. *Unlink your horses!*
153. *Files, to the right double!*
154. *March!*
155. *Fix your links!*
156. *Make ready to mount!*
157. *Take your left stirrup!*
158. *Mount!*

Note. The Officers perform the above three words of command with the men.

159. *Handle your slings!*
160. *Poise your firelocks!*
161. *Advance your firelocks!*

Bring down your butt to your right thigh, the barrel towards you, the muzzle sloping a little forwards.

162. *Return your firelocks!*

Throw back your firelock to your right side, seizing it at the same time with your left hand a little above the lock. Tell 1, 2, and bring your right hand between your body and the firelock, and seize it immediately above the left hand. Tell 1, 2, quit your firelock with the left hand, thrusting it down briskly by your right side with the right hand, and put it in the bucket.

The Evolutions are the same as before explained for the Horse.

CHAP. XXII.

Part of the Exercise for the Troops of Light Dragoons.

THE Light Troops of Dragoons are to form separately on the right of the Regiment; and, when dismounted, are to be drawn up as the Company of Grenadiers of the infantry; covering the flanks of the Battalion when it is to go through the firings as in the foot; the Platoon Exercise is therefore to be the same as that of the Dragoons.—The dismounting and mounting is to be done by the usual words of command given to the Regiment; the Major giving time to the Light Troops, when they are exercised with the Regiment, to do the motions which are peculiar to them. The following are the words of command for the Light Troops when exercised separately from the Regiment, with the number of motions contained in each, and an explanation of such, as are different from what is practised by the other Troops of Dragoons.

Words of command for the carbine when the Light Troop is exercised separately from the Regiment.

Return your swords!

Files, to the right double!

March! (pull off gloves)

Note. When no explanations follow a word of command, the motions contained in it, are to be done as is above directed in the Horse Exercise.

Make ready your carbines! 4 motions.

At the fourth motion of the firelock Exercise, drop your carbine by your right side.

Throw back your carbines! 3 motions.

1st, Seize the carbine with a full hand a little above the guard.

2d, Bring the butt and muzzle in a line, by raising your wrist and elbow square, pointing your muzzle directly forwards.

3d, Throw it cross behind with life, the butt to the near side, quitting your hand.

Make ready your links!

The links are briskly unfixed, and dropped on the horse's neck; and the bridle reins immediately taken up, without any distinct motions.

Quit your right stirrups!
Dismount!

Immediately after the last motion, take the sword and scabbard out of the sword belt, by clearing the hook with the thumb and fore-finger of the right hand from the loop of the belt; with the same hand seize the sword below the hilt, the thumb downwards, and the back of the hand turned outwards; pull the sword and scabbard briskly out of the belt, turning the palm of the hand upwards, which will bring the point of the sword to the near cloak strap up; fix it in the loop with the right hand, and strap the hilt betwixt the holsters with the carbine strap.

Face to the left!
Files that doub.ed, as you were!
March!
To the right about!
Link your horses!
To the right about!

Move

Chap. XXII.   *Military Discipline.*

Move clear of your horses!
March!

Unspring your carbines!   3 motions.

1ft, Seize your carbine with the right hand at the loweft pipe, the palm upon the barrel.

2d, Bring up your carbine upon the right fide, feizing it with the left hand above the feather fpring.

3d, Bring your right hand to the fwivel, and loofen it from the ring, holding the carbine with the left hand.

Poife your carbines!   3 motions.

1ft, Throw back the fwivel with your right hand.

2d, Seize the carbine under the hammer with your right hand, bringing it before your breaft with both hands.

3d, Come to a poife.

Shoulder your carbines!

After the Dragoons have gone through the Foot Exercife, and are ordered back to their horfes.

Poife your carbines!
Spring your carbines!   2 motions.

1ft, Throw back your carbine to the right fide as for ordering, feizing it with the left hand above the feather fpring.

2d, Seize your fwivel with your right hand, and put it in the ring.

Sling your carbines!   3 motions.

1ft, Seize your carbine with the palm of your right hand upon the rammer between the fecond and third pipe.

2d, Throw back the muzzle with your right hand, and feize your fhoulder belt with your left; your thumb under, and fingers above.

3d, Draw your fhoulder belt round with your left hand, throwing behind, with your right hand, the carbine.

To the right-about!
Move to your horses!
March!
Unlink your horses!
Files, to the right-about!
March!
Fix your links! 2 motions.

1st, Step with the left heel, in a line with the horses fore feet.

2d, Face to the left on the left heel, and fix your links, putting your swords into the sword belts as quick as possible.

Make ready to mount!
Take your left stirrups! 2 motions.

1st, Seize the stirrup, taking a large step to the rear with the right foot.

2d, Put the foot in the stirrup, bringing the right hand back to the carbine.

Mount!

As soon as mounted, bring the carbine over the sword with the left hand, holding it there, back of the hand to the rear.

Handle your carbines! 3 motions.

1st, Bring up the carbine with the left hand level over the cloak, seizing it at the last pipe with the right hand, back of the hand to the rear.

2d, Quitting the left hand, bring the carbine upright with the right hand, the guard opposite to the outside of the right thigh.

3d, Let the carbine drop by the right side behind the leg, catching it again with the right hand as it drops, to ease the swivel, quitting it immediately after.

Return your carbines! 3 motions.

1ſt, Seize the carbine with your right hand, the palm upon the barrel.

2d, Bring it forwards, the butt to the ſide of the holſter, turning the barrel upwards.

3d, Strike it ſtrong into the bucket.

Evolutions for the Light Dragoons.

I. Fire carbines to the front by whole ranks, wheeling outwards by two diviſions to the rear.

Make ready, preſent and fire to each rank, each rank wheeling off immediately after firing, with recovered carbines (*viz.* as in the poſition of the laſt motion of making ready) to form behind the rear rank and load.

II. Fire carbines to the front by half ranks, each half rank wheeling outwards to the rear, after firing.

III. Fire carbines to the front by whole ranks, wheeling by fours to the right, to form before the right interval of their own Squadron.

IV. Fire carbines by whole ranks, file from right and left to fire piſtols on the enemy's flanks, returning immediately to the right-about.

V. Fire carbines by whole ranks, file to fire piſtols on one flank, charging afterwards with ſwords drawn.

The front and center rank to charge the ſame flank they fire on; the one the right, the other the left flank.

The rear rank to charge the rear, after firing. Each rank, after charging, files off from the flank that led, to join the diviſion.

VI. File by half ranks from right to left acroſs the enemy's front, firing carbines.

The

The front and center, after firing carbines, will immediately return to the rear of their division.

The rear rank files entire; and after firing, files round the enemy's rear, forms, fires pistols, and follows with swords drawn.

VII. File by whole ranks, fire carbines on the flanks, pistols in the rear, and charge the other flank with swords drawn.

The front rank files from the right.

The center from the left.

The rear from the right.—And, after firing carbines, forms in the enemies rear, fires pistols, and follows with swords drawn.

VIII. Street firing, advancing.

IX. Street firing, retreating.

X. Incline in the front of the Squadron, and, when there, form a rank entire to the front, fire carbines, file from right and left, fire pistols, returning to the right-about, immediately.

XI. Fire carbines and pistols in a rank entire, following with swords drawn.

XII. When the heavy Squadron charges, the division of Light Dragoons to form two deep on its right (at the time the Squadrons shock) ready to fire, and when the enemies Squadron retreats, they will form a rank entire to the front, fire pistols, and pursue with swords drawn, forming in the rear of the heavy Squadron, at assembly-beating.

XIII. And if ordered to assist the heavy Squadron in the charge, and support in their retreat, they will form on the flank, at the charge beating, and fire carbines by whole ranks, as the enemy is about joining the Squadron.

And the Squadron being obliged to retreat, they will fire piftols on the enemies flank and rear, charge in on both immediately after, and form in the rear of their own Squadron at the word, to the right-about.

XIV. When the detachments of light Dragoons are advanced before their refpective Squadrons of the line, as foon as the lines advance to charge, the detachments muft immediately file by ranks from the right, and form in the rear.

XV. Charge two deep, and detach flanks to purfue.

XVI. Retreat, and fire flanks.

XVII. Charge, and the whole purfue.

CHAP. XXIII.

ARTICLE I.

Rules for the reception of a General Officer, who comes to review the Regiment, or see the Exercise performed.

WHEN the General comes in view, the Field-Officers are to post themselves on the front of their Squadrons, a horse's length before the rank of Officers, in a direct line with the Standards. The Colonel on the right, the Lieutenant-Colonel on the left Squadron, and the Major on the center: But if the Regiment consists only of two Squadrons, the Major is to post himself a little to the left of the Colonel on the first Squadron, and about half a horse's length further back. The Cornets are at the same time to take the Standards, and to fall into the center of the front rank, and dress with the men, unless they are to salute, and then they are to be half a horse's length before them, for the easier performing it.

As soon as the General comes near the right flank of the first Squadron, the Colonel is to order the swords to be drawn, at which time the Trumpets are to sound, or Drums beat, a *March*; and, as the General passes along the front, the Officers are to salute him with their swords. If he is to be saluted with the Standard, the Cornet is to do it, when the Colonel or Commanding Officer of the Squadron salutes, and to keep it down till he is past him.

When he approaches the center Squadron, it is to do the same as the first, and so with the left Squadron, they being not to draw their swords, or beat a *March* till he comes

comes near them, each Squadron being then a distinct body.

When the General has reviewed them standing, it is usual, if time permits, to see the Exercise performed; after which, to review them marching, either by entire Squadrons, single Troops, by fours, or by ranking off singly, as he shall be pleased to order.

If by entire Squadrons (as soon as the General has taken his place) the center and rear ranks are to move up to order, and the Kettle-Drummer and Trumpets or Drums, are to place themselves in the front of the Colonel, or Officer commanding the Squadron, and to march in that order by the General, leaving but two or three paces between the right flank of the Squadron and him. When the commanding Officer comes within a proper distance of the General, he is to salute him; and when the rank of Officers comes upon the same ground they are to salute together; as is also the Cornet with the Standard (if ordered to salute) when he comes to the same place. When the Squadrons have passed the General, they are to wheel and march back, and draw up on their former ground.

If by single Troops, the ranks are to close to order, the Officers and Trumpets, or Drums, ordered to their Troops, and the Standards to the center of the front rank of the eldest Troop in each Squadron. When this is done, they are to march in Squadron, till they come within twenty or thirty paces of the General, and then the Troops on the left and center halt, but the Troop on the right marches on, and the other Troops (as soon as it is passed them) passage to the right, till they come to the same ground, and then march forward, keeping directly in the rear of one another. In this position, the Captain is to be in the center, the Lieutenant on his right, and the Cornet on his left, but the Captain half a horse's length before them. In those Troops that have the Standards, the Cornets are to remain in the front rank of men, and march and dress with them, unless they are to salute, and then they are to be half a horse's length advanced before them for the conveniency of performing it. In this case, the Lieutenant is to be on the left of the Captain, but not so far advanced. The Trumpets or Drums are to march before the Captains, and the Quarter-Masters in the rear of the Troops.

If they are to march thus by the General, the Officers are to salute together; but as it is more usual when they are in single Troops, to rank off singly, I will set down the directions accordingly.

The Squadrons being reduced into Troops, as soon as the first Troop comes within ten paces of the General, it is to halt, but the Trumpets or Drums keep moving on, and the Captain follows, then the Lieutenant, and next the Cornet (unless he carries the Standard, and then he remains in the front rank, till the men on his right are marched off before him) then the right hand man of the front rank follows him, and the rest of that rank passages to the right, and when the men come upon the ground where the right hand man stood, they are to move out, and march directly in the rear of one another. When the front rank is marched off, the center moves up into their ground, and marches off singly as the first did, and the rear rank moves up at the same time to the ground on which the center rank stood, and then halt, till that rank is marched off, and then it moves up, and ranks off singly, as the others.

The other Troops are to pass by the General in the same manner, and not rank off till they come upon the same ground.

When the men have got about twenty paces beyond the General, they are to rank up, and the Troop is to march about twenty paces more, when it is to halt, till the other Troops join; and when the Squadron is again formed, it is to march and draw up on the former ground.

The other Squadrons are to follow the same directions. When the Squadrons are to march by fours, the ranks are to be closed to order, and the Officers ordered to the front of the first four men on the right, the Captains

placing

Chap. XXIII.   *Military Discipline.*   399

placing themselves by seniority in one rank, the Lieutenants in another in the rear of the Captains, and the Cornets in the rear of the Lieutenants, (except the Cornet that carries the Standard, who remains in the front rank of men, and is to march in the center of the rank of fours he is then placed in) and the Quarter-Masters to march in the rear of their respective Squadrons. The Commanding Officer of each Squadron is to march before the Captains, and the Kettle-Drummer and Trumpets, or Drums, before him. All things being thus disposed, the Commanding Officer is to give the word, *March!* at which the Officers, and the first four men on the right of the front rank, are to march straight forward, and the rest of the front rank are to wheel to the right by fours, and march till they come to the ground where the first rank of fours stood, and then wheel to the left, taking care to keep their proper distance, and to cover the rank before them.

As soon as the last four men of the front rank are come to the ground where the right stood, the center rank is to move up to the ground of the front rank, and the rear rank up to the ground of the center, and then halt till the center rank is marched off, which they are to do in the same manner as the front rank did. The same rule is to be observed by the rear rank as soon as the center is marched off.

Each rank of Officers is to salute as they come on the ground, where the Commanding Officer saluted.

When they are got to a proper distance beyond the General, they should then form into Squadron, either by ranking up, or each rank forming at once a rank entire to the left.

Ranking up, is when the second rank of fours draws up on the left of the first, the third on the left of the second, and so on through those of the front rank. The first rank of fours of the center rank draws up in the rear of the right of the front rank, leaving a proper interval between them, and the rest of that rank draws up on their left, at those of the front rank did. The rear rank observes the same rules in drawing up in the rear

F f 2

of the center, as the center did in regard to the front rank. As they are drawing up, the Officers are to divide themselves on the front of the Squadron.

If you would form each rank into a rank entire at once, you must order all the ranks of fours of the front rank to wheel at once to the left, which forms them into a rank entire, and consequently forms the front of the Squadron. As soon as that is done, the Officers are immediately to divide themselves on the front, and to march forward, that the center rank may march in the rear of the front, and form a rank entire, which it is to do when the first four comes near the right of the front rank, without any word of command, and then march forward, that the rear rank may do the same. After this the Squadrons are to draw up on their former ground with their ranks at open order, and to keep their swords drawn till the General quits the field.

FINIS.

www.ingramcontent.com/pod-product-compliance
Lightning Source LLC
Chambersburg PA
CBHW071309150426
43191CB00007B/556